Inventor 2020 for Novices

(Learn By Doing)

CADSoft Technologies

Inventor for Novices:
Learn By Doing

CADSoft Technologies

CADSoft Technologies

ISBN: 978-1-64057-068-9

DEDICATION

To students, who are dedicated to learning new technologies

THANKS

To the faculty and employees of CADSoft Technologies
for their cooperation

Training Programs offered by CADSoft Technologies

CADSoft Technologies provides quality training on various software packages including Computer Aided Design, Manufacturing and Engineering (CAD/CAM/CAE), Web and Programming Languages, animation, architecture, and GIS. The training is delivered at its training centers or online at any time, place, and pace to individuals, corporates as well as the students of colleges, universities, and CAD/CAM/CAE training centers. The main features of this program are:

Training in Classroom Settings and Online Training

A team of highly experienced instructors and qualified engineers at CADSoft Technologies conduct the classes in classroom environment at its centers. This team has authored several textbooks that are rated "one of the best" in their categories and are used in various colleges, universities, and training centers in North America, Europe, and in other parts of the world.

CADSoft Technologies with its cost effective and time saving initiative strives to deliver online training in the comfort of your home or work place, thereby relieving you from the hassles of traveling to training centers. CADSoft Technologies strives to be the best training institute in USA. CADSoft Technologies has developed career oriented programs by following state-of-the-art teaching and learning methodologies.

Courses Offered by CADSoft Technologies

CADSoft Technologies conducts classes on the following software packages:

CAD/CAM/CAE: *CATIA, Pro/ENGINEER Wildfire, Creo Parametric, Creo Direct, SOLIDWORKS, Autodesk Inventor, Solid Edge, NX, AutoCAD, AutoCAD LT, AutoCAD Plant 3D, Customizing AutoCAD, SolidCAM, NX CAM, NX Mold, Creo Mold, Alias Design, Alias Automotive, Ansys Fluent, and ANSYS*

Architecture and GIS: *Autodesk Revit Architecture, AutoCAD Civil 3D, Autodesk Revit Structure, AutoCAD Map 3D, Revit MEP, Navisworks, Primavera, and Bentley STAAD Pro, MS Project, MX Road, ArcGIS, and Raster Design*

Animation and VFX: *Autodesk 3ds Max, Chaosgroup VRay, Autodesk 3ds Max Design, Autodesk Maya, Blackmagic Design Fusion Studio, Adobe Premiere, Adobe Photoshop, Adobe Indesign, Adobe Illustrator, Corel Graphic Design CorelDraw, The Foundry NukeX, and MAXON CINEMA 4D*

Web and Programming: *C/C++, HTML5/CSS3, JavaScript, jQuery, Bootstrap, PHP, MySQL, Dreamweaver, VB.NET, Oracle, AJAX, and Java*

Personality Development*: Personality Development and Engineering Ethics/Soft Skills Course*

Table of Contents

Chapter 4: Editing, Extruding, and Revolving the Sketches

Chapter 5: Other Sketching and Modeling Options

Chapter 6: Advanced Modeling Tools-I

Chapter 7: Editing Features and Adding Automatic Dimensions to Sketches

Chapter 8: Advanced Modeling Tools-II

Chapter 9: Assembly Modeling-I

Chapter 10: Assembly Modeling-II

Chapter 11: Working with Drawing Views-I

Chapter 12: Working with Drawing Views-II

Preface

Autodesk Inventor Professional 2020

Autodesk Inventor, developed by Autodesk Inc., is one of the world's fastest growing solid modeling software. It is a parametric feature-based solid modeling tool that not only unites the 3D parametric features with 2D tools but also addresses every design-through-manufacturing process. The adaptive technology of this solid modeling tool allows you to handle extremely large assemblies with tremendous ease. Based mainly on the feedback of the users of solid modeling, this tool is known to be remarkably user-friendly and allows you to be productive from day one.

This solid modeling tool allows you to easily import the AutoCAD, AutoCAD Mechanical, Mechanical Desktop, and other related CAD files with an amazing compatibility. Moreover, the parametric features and assembly parameters are retained when you import the Mechanical Desktop files in Autodesk Inventor.

The drawing views that can be generated using this tool include orthographic view, isometric view, auxiliary view, section view, detailed view, and so on. You can use predefined drawing standard files for generating the drawing views. Moreover, you can retrieve the model dimensions or add reference dimensions to the drawing views whenever you want. The bidirectional associative nature of this software ensures that any modification made in the model is automatically reflected in the drawing views. Similarly, any modifications made in the dimensions in the drawing views are automatically reflected in the model.

Inventor 2020 for Novices: Learn By Doing textbook is written with the intention of helping the readers effectively use the Autodesk Inventor Professional 2020 solid modeling tool. The mechanical engineering industry examples and tutorials are used in this textbook to ensure that the users can relate the knowledge of this book with the actual mechanical industry designs. The salient features of this textbook are as follows:

- **Tutorial Approach**

  The author has adopted the tutorial point-of-view and the learn-by-doing approach throughout the textbook. This approach guides the users through the process of creating the models in the tutorials.

- **Real-World Projects as Tutorials**

  The author has used about 28 real-world mechanical engineering projects as tutorials in this book. This enables the readers to relate these tutorials to the real-world models in the mechanical engineering industry. In addition, there are about 40 exercises that are also based on the real-world mechanical engineering projects.

- **Coverage of All Basic Autodesk Inventor Modules**

  Basic modules of Autodesk Inventor are covered in this book including Assemblies and Drwawings.

- **Tips and Notes**

  Additional information on various topics is provided to the users in the form of tips and notes.

- **Learning Objectives**

  The first page of every chapter introduces in brief the topics that are covered in that chapter. This helps the users to easily refer to a topic.

- **Exercises**

  Exercises are given at the end of the chapter and they can be used by the instructors as test questions and exercises.

  These exercises are used as reference in tutorials. Make sure you perform all the exercises as output of some of these exercises are used as reference in this book.

Symbols Used in the Textbook

Note

The author has provided additional information to the users about the topic being discussed in the form of notes.

Tip

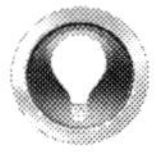

Special information and techniques are provided in the form of tips that helps in increasing the efficiency of the users.

Formatting Conventions Used in the Textbook

Please refer to the following list for the formatting conventions used in this textbook.

- Names of tools, buttons, options, panels, tabs, and Ribbon are written in boldface.
  Example: The **Extrude** tool, the **Finish Sketch** button, the **Modify** panel, the **Sketch** tab, and so on.

- Names of dialog boxes, drop-downs, drop-down lists, list boxes, areas, edit boxes, check boxes, and radio buttons are written in boldface.
  Example: The **Revolve** dialog box, the **Start 2D Sketch** drop-down of **Sketch** panel in the **Model** tab, the **Placement** drop-down in the **Hole** dialog box, the **Distance** edit box of the **Extrude** dialog box, the **Extended Profile** check box in the **Rib** dialog box, the **Drilled** radio button in the **Hole** dialog box, and so on.

- Values entered in edit boxes are written in boldface.
  Example: Enter **5** in the **Radius** edit box.

- Names and paths of the files are written in italics.
  Example: *C:\Inventor2020\c03*, *c03tut03.prt*, and so on

- The methods of invoking a tool/option from the **Ribbon**, **Quick Access Toolbar**, **Application Menu** are enclosed in a shaded box.

  Ribbon: Get Started > Launch > New
  Quick Access Toolbar: New
  Application Menu: New

Naming Conventions Used in the Textbook

Tool

If you click on an item in a toolbar or a panel of the **Ribbon** and a command is invoked to create/edit an object or perform some action, then that item is termed as **tool**.

For example:
To Create: **Line** tool, **Dimension** tool, **Extrude** tool
To Edit: **Fillet** tool, **Draft** tool, **Trim Surface** tool
Action: **Zoom All** tool, **Pan** tool, **Copy Object** tool

If you click on an item in a toolbar or a panel of the **Ribbon** and a dialog box is invoked wherein you can set the properties to create/edit an object, then that item is also termed as **tool**, refer to Figure 1.

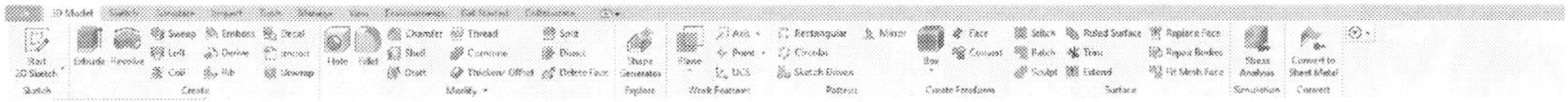

Figure 1** Various tools in the **Ribbon

For example:
To Create: **Create iPart** tool, **Parameters** tool, **Create** tool
To Edit: **Styles Editor** tool, **Document Settings** tool

Button

The item in a dialog box that has a 3d shape like a button is termed as **Button**. For example, **OK** button, **Cancel** button, **Apply** button, and so on.

Dialog Box

In this textbook, different terms are used for referring to the components of a dialog box. Refer to Figure 2 for the terminology used.

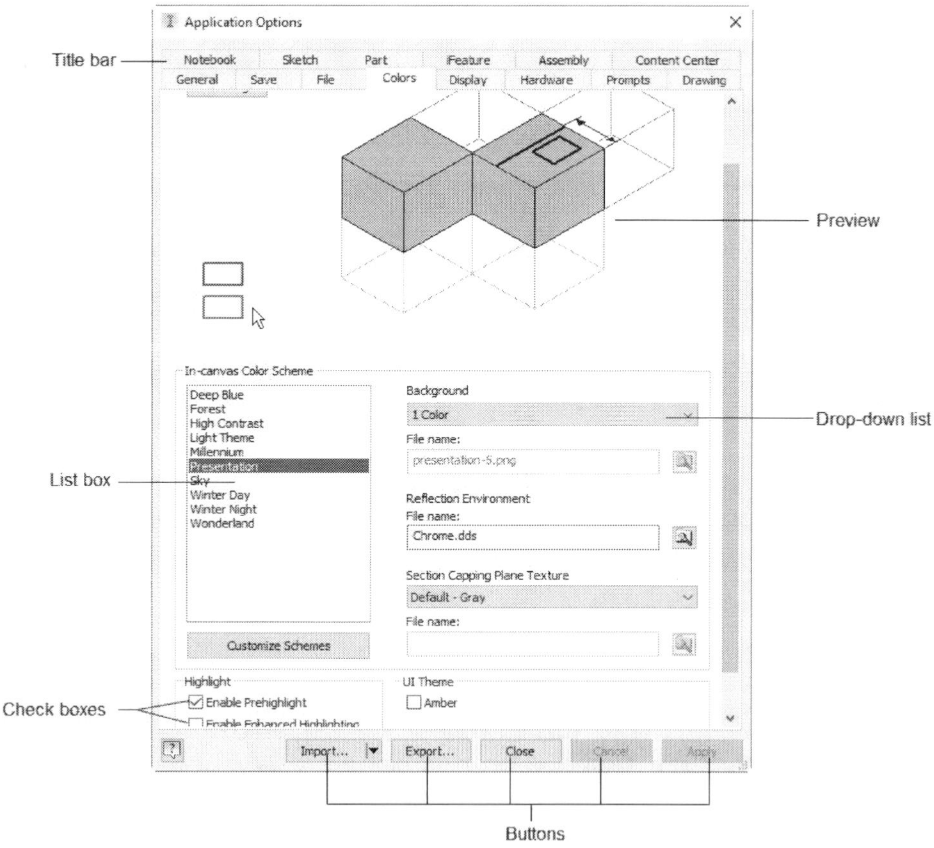

Figure 2 The components of a dialog box

Drop-down

A drop-down is one in which a set of common tools are grouped together. You can identify a drop-down with a down arrow on it. These drop-downs are given a name based on the tools grouped in them. For example, **Arc** drop-down, **Fillet/Chamfer** drop-down, **Work Axis** drop-down, and so on; refer to Figure 3.

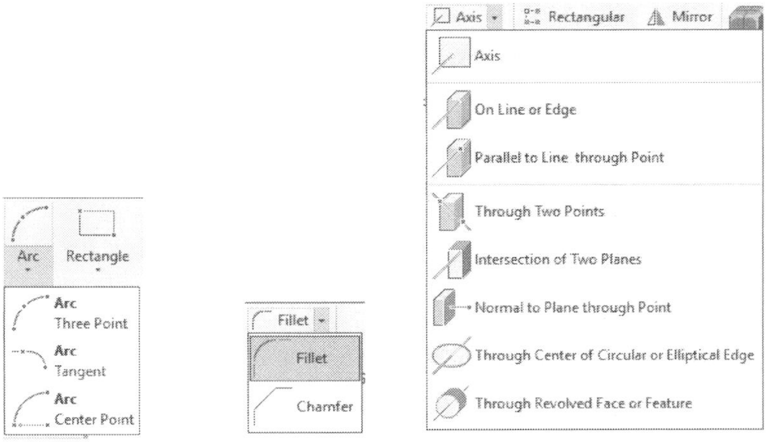

***Figure 3** The **Arc**, **Fillet/Chamfer**, and **Work Axis** drop-downs*

Drop-down List

A drop-down list is the one in which a set of options are grouped together. You can set various parameters using these options. You can identify a drop-down list with a down arrow on it. For example, **Extents** drop-down list, **Color Override** drop-down list, and so on, refer to Figure 4.

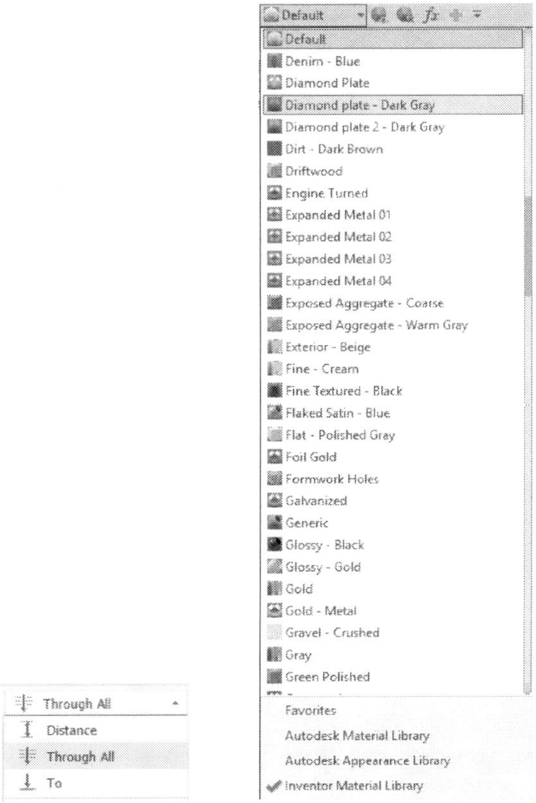

***Figure 4** The **Termination** and **Color Override** drop-down lists*

Options

Options are the items that are available in shortcut menu, Marking Menu, drop-down list, dialog boxes, and so on. For example, choose the **New Sketch** option from the Marking Menu displayed on right-clicking in the drawing area; choose the **Background Image** option from the **Background** drop-down list; choose the **Front** option from the **Orientation** area, refer to Figure 5.

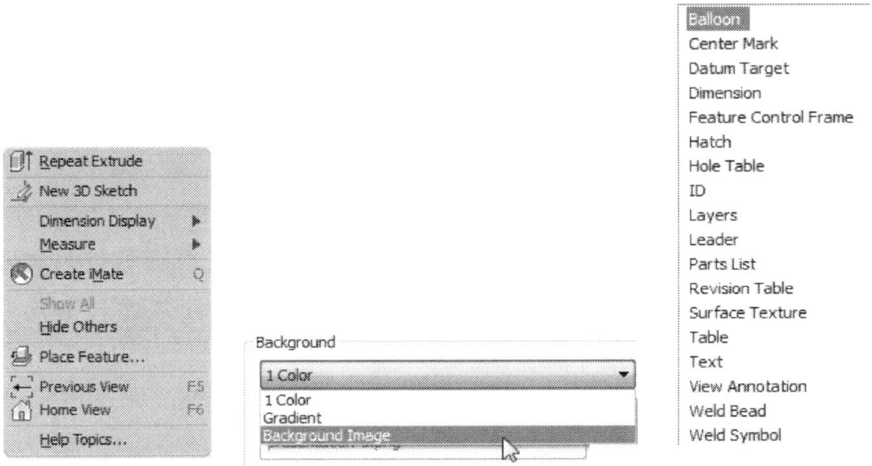

***Figure 5** Options in the shortcut menu, **Background** drop-down list, and the **Style Type** area*

Chapter 1

Introduction

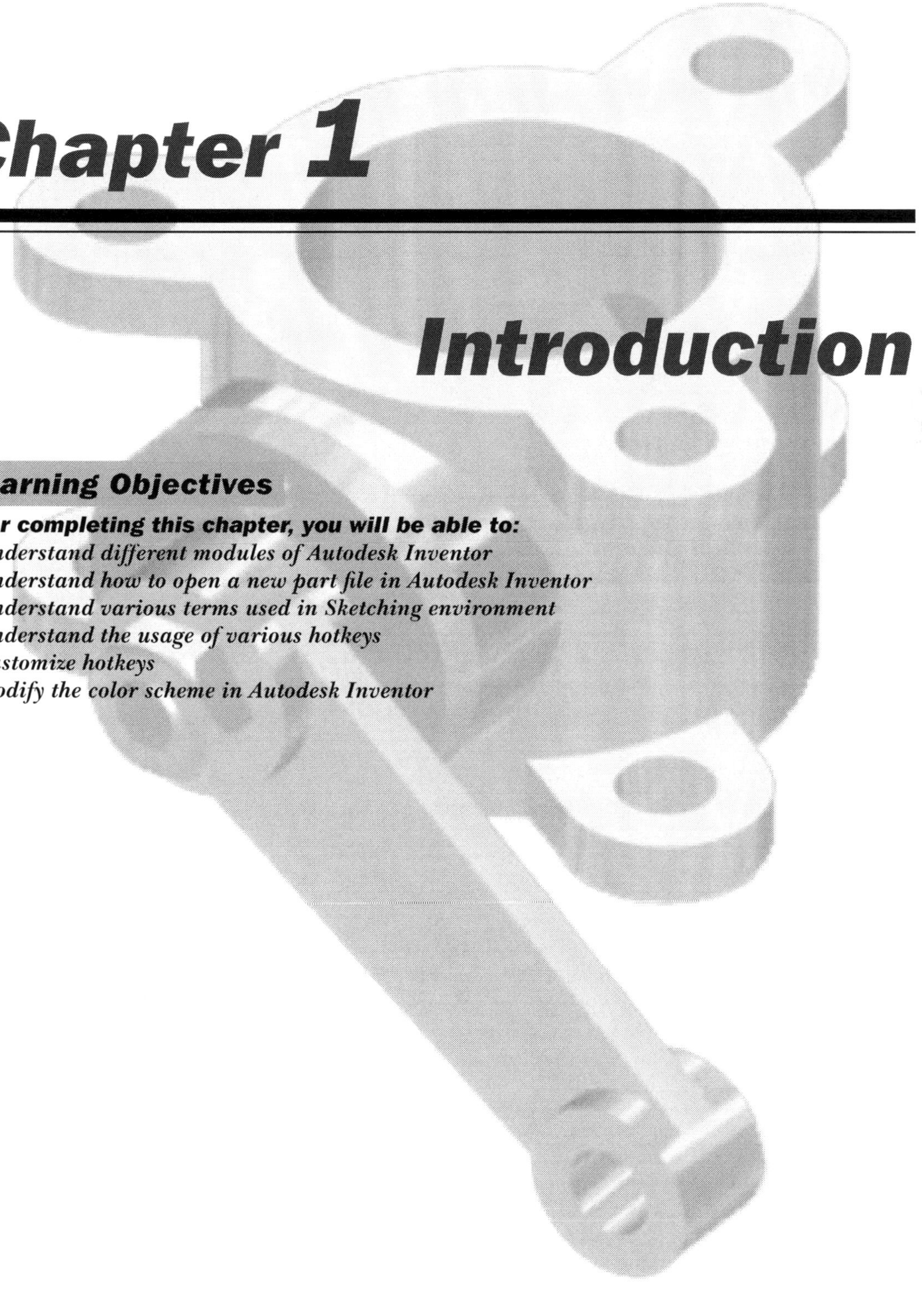

Learning Objectives

After completing this chapter, you will be able to:

- *Understand different modules of Autodesk Inventor*
- *Understand how to open a new part file in Autodesk Inventor*
- *Understand various terms used in Sketching environment*
- *Understand the usage of various hotkeys*
- *Customize hotkeys*
- *Modify the color scheme in Autodesk Inventor*

INTRODUCTION TO Autodesk Inventor 2020

Welcome to the world of Autodesk Inventor. If you are new to the world of three-dimensional (3D) design, then you have joined hands with thousands of people worldwide who are already working with 3D designs. If you are already using any other solid modeling tool, you will find this solid modeling tool more adaptive to your use. You will find a tremendous reduction in the time taken to complete a design using this solid modeling tool.

Autodesk Inventor is a parametric and feature-based solid modeling tool. It allows you to convert basic two-dimensional (2D) sketch into a solid model using very simple but highly effective modeling options. This solid modeling tool does not restrict its capabilities to the 3D solid output but also extends them to the bidirectional associative drafting. This means that you only need to create the solid model. Its documentation, in the form of the drawing views, is easily done by this software package itself. You just need to specify the required view. This solid modeling tool can be specially used at places where the concept of "collaborative engineering" is brought into use. Collaborative engineering is a concept that allows more than one user to work on the same design at the same time. This solid modeling package allows more than one user to work simultaneously on the same design.

As a product of Autodesk, this software package allows you to directly open the drawings of the other Autodesk software like AutoCAD, Mechanical Desktop, AutoCAD LT, and so on. This interface is not restricted to the Autodesk software only. You can easily import and export the drawings from this software package to any other software package and vice versa.

To reduce the complicacies of design, this software package provides various design environments. This helps you capture the design intent easily by individually incorporating the intelligence of each of the design environments into the design. The design environments that are available in this solid modeling tool are discussed next.

Part Module

This is a parametric and feature-based solid modeling environment and is used to create solid models. The sketches for the models are also drawn in this environment. All applicable constraints are automatically applied to a sketch while drawing. You do not need to invoke an extra command to apply them. Once the basic sketches are drawn, you can convert them into solid models using simple but highly effective modeling options. One of the major advantages of using Autodesk Inventor is the availability of the Design Doctor. The Design Doctor is used to calculate and describe errors, if any, in the design. You are also provided with remedy for removing errors such that the sketches can be converted into features. The complicated features can be captured from this module and can later be used in other parts. This reduces the time taken to create the designer model. These features can be created using the same principles as those for creating solid models.

Assembly Module

This module helps you create the assemblies by assembling multiple components using assembly constraints. This module supports both the bottom-up approach as well as the top-down approach of creating assemblies. This means that you can insert external components into the **Assembly** module or create the components in the **Assembly** module itself. You are allowed to assemble the components using the smart assembly constraints and joints. All the assembly constraints and joints can be added using a single dialog box. You can even preview the components before they are actually assembled. This solid modeling tool supports the concept of making a part or a feature in the part adaptive. An adaptive feature or a part is the one that can change its actual dimensions based upon the need of the environment.

Presentation Module

A major drawback of most solid modeling tools is their limitation in displaying the working of an assembly. The most important question asked by customers in today's world is how to show the working of any assembly. Most of the solid modeling tools do not have an answer to this question. This is because they do not have proper tools to display an assembly in motion. As a result, the designers cannot show the working of the assemblies to their clients or they have to take the help of some other animation software packages. However, this software package provides a module called the **Presentation** module using which you can animate the assemblies created in the **Assembly** module and view their working. You can also view any interference during the operation of the assembly. The assemblies can be animated using easy steps.

Drawing Module

This module is used for the documentation of the parts or assemblies in the form of drawing views. You can also create drawing views of the presentation created in the **Presentation** module. All parametric dimensions added to the components in the **Part** module during the creation of the parts are displayed in the drawing views in this module.

Sheet Metal Module

This module is used to create a sheet metal component. You can draw the sketch of the base sheet in this Sketching environment and then proceed to the sheet metal module to convert it into a sheet metal component.

Mold Design Module

This module is used to create mold design by integrated mold functionality and content libraries using the intelligent tools and catalogs provided in mold design module. In this module, you can quickly generate accurate mold design directly from digital prototypes.

GETTING STARTED WITH Autodesk Inventor

Install Autodesk Inventor on your system; a shortcut icon of Autodesk Inventor Professional 2020 will automatically be created on the desktop. Double-click on this icon to start Autodesk Inventor.

When Autodesk Inventor is started for the first time, the system prepares itself by loading all the required files and then the **Welcome to Inventor 2020** window gets displayed, as shown in Figure 1-1.

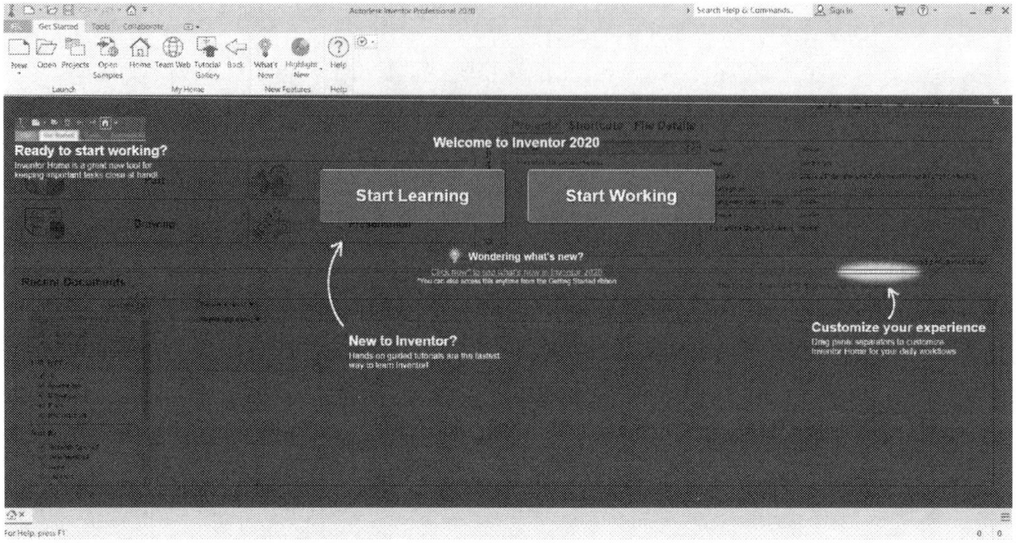

*Figure 1-1 The **Welcome to Inventor 2020** window*

On choosing the **Start Learning** button from this window, the **Tutorials** window will be displayed, as shown in Figure 1-2.

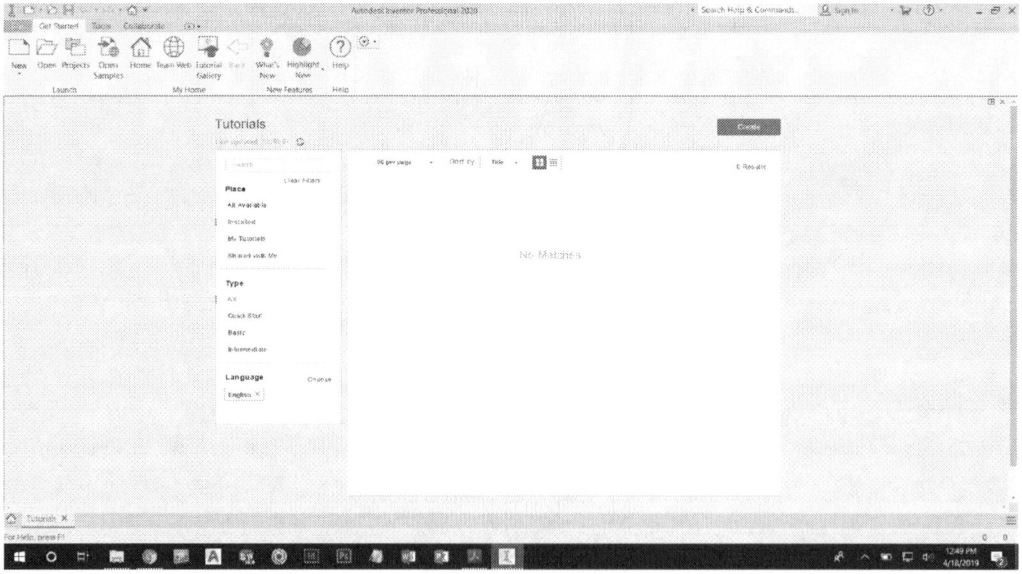

*Figure 1-2 The **Tutorials** window*

In this window, you can see the basic and intermediate level tutorials uploaded by Autodesk. Also, you can create tutorials and upload them in the cloud by choosing **CREATE** button on right side of this window. On choosing the **Start Working** button from the **Welcome to Inventor 2020** window, the initial interface of Autodesk Inventor Professional 2020 will be displayed, as shown in Figure 1-3.

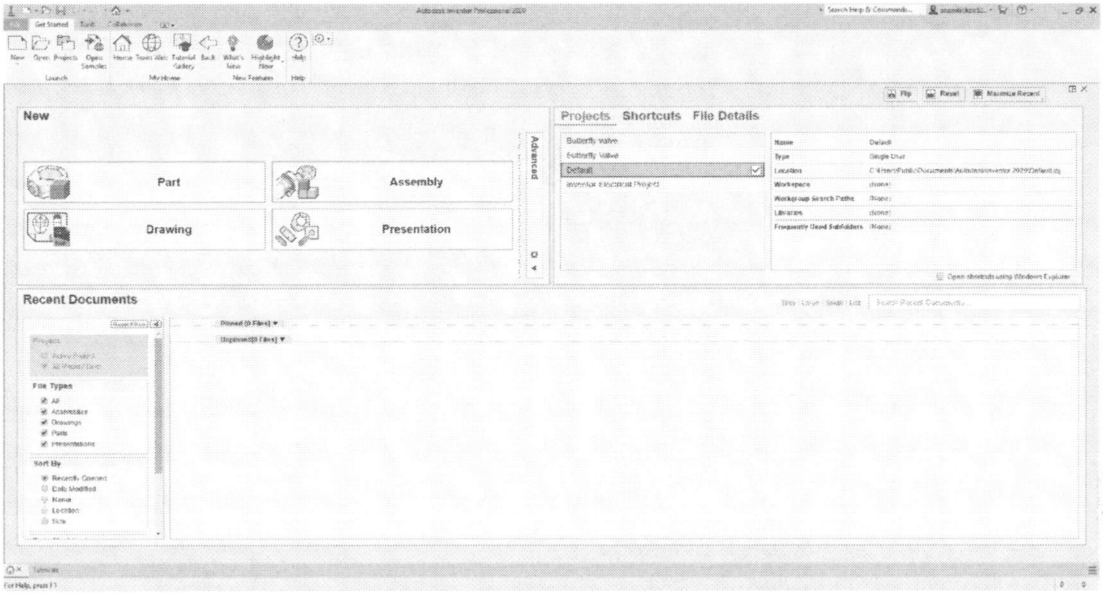

Figure 1-3 *Initial interface of Autodesk Inventor Professional 2020*

It is evident from Figure 1-4 that the interface of Autodesk Inventor is quite user-friendly. Apart from the components shown in Figure 1-4, you are also provided with various shortcut menus which are displayed on right-clicking in the drawing area. The type of the shortcut menu and its options depend on where or when you are trying to access the menu. For example, when you are inside any command, the options displayed in the shortcut menu will be different from the options displayed when you are not inside any command. The different types of shortcut menus will be discussed when they are used in the textbook.

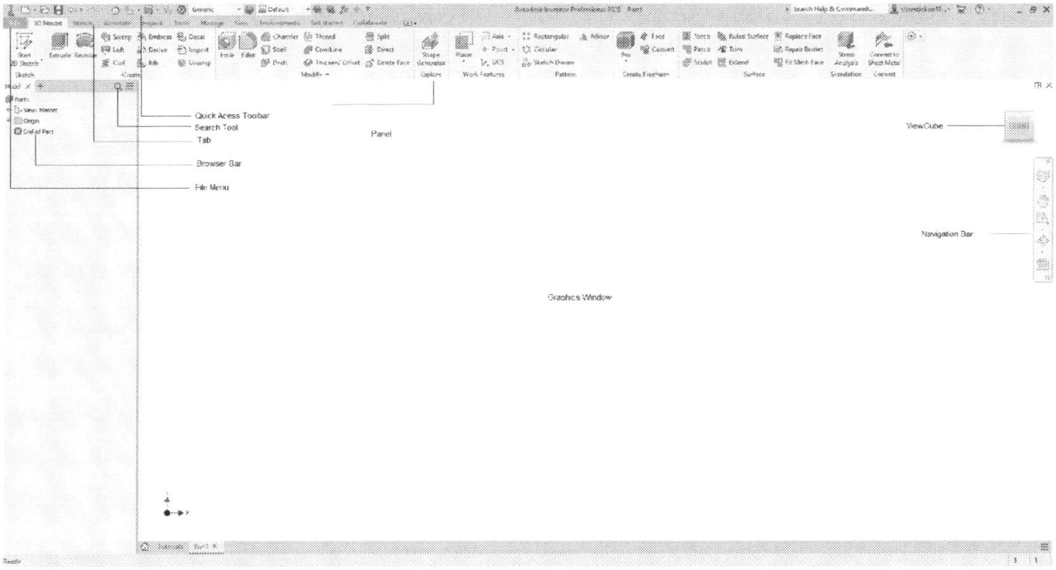

Figure 1-4 *Components of Autodesk Inventor interface*

HOTKEYS

As mentioned earlier, there is no command prompt in Autodesk Inventor. However, you can use the keys on the keyboard to invoke some tools. The keys that can be used to invoke the tools are called hotkeys. Remember that the working of the hotkeys will be different for different environments. The use of hotkeys in different environments is given next.

Part Module

The hotkeys that can be used in the **Part** module and their functions are given next.

Hotkey	Function
E	Invokes the **Extrude** tool
R	Invokes the **Revolve** tool
H	Invokes the **Hole** tool
CTRL+SHIFT+L	Invokes the **Loft** tool
CTRL+SHIFT+S	Invokes the **Sweep** tool
F	Invokes the **Fillet** tool
CTRL+SHIFT+K	Invokes the **Chamfer** tool
CTRL+SHIFT+M	Invokes the **Mirror** tool
CTRL+SHIFT+R	Invokes the **Rectangular Pattern** tool
CTRL+SHIFT+O	Invokes the **Circular Pattern** tool
F6	Invokes the **Home view**
]	Invokes the **Work Plane** tool
/	Invokes the **Work Axis** tool
.	Invokes the **Work Point** tool
CTRL+W	Invokes the **SteeringWheels**

The following hotkeys are used in the Sketching environment:

Hotkey	Function
L	Invokes the **Line** tool
D	Invokes the **Dimension** tool
X	Invokes the **Trim** tool
F7	Invokes the **Slice Graphics** tool
F8	Displays all constraints
F9	Hides all constraints

Assembly Module

In addition to the hotkeys of the part modeling tool, the following hot keys can also be used in the **Assembly** module:

Hotkey	Function
P	Invokes the **Place** tool
N	Invokes the **Create** tool
C	Invokes the **Constrain** tool
V	Invokes the **Free Move** tool
G	Invokes the **Free Rotate** tool

Drawing Module

The hotkeys that can be used in the **Drawing** module are given next.

Hotkey	Function
B	Invokes the **Balloon** tool
D	Invokes the **Dimension** tool
O	Invokes the **Ordinate Set** tool
F	Invokes the **Feature Control Frame** tool

In addition to these keys, you can also use some other keys for the ease of designing. Note that you will have to hold some of these keys down and use them in combination with the pointing device. These hotkeys are given next.

Hotkey	Function
F1	Invokes the **Help** command
F2	Invokes the **Pan** tool
F3	Invokes the **Zoom** tool
F4	Invokes the **Free Orbit** tool
F5	Previous view
SHIFT+F5	Next view
ESC	Aborts the current command
SPACEBAR	Invokes the recently used tool
T (In **Presentation** module)	Invokes the **Tweak Components** tool

Customizing Hotkeys

You can customize the settings of hotkeys. To do so, choose the **Customize** tool from the **Options** panel of the **Tools** tab in the **Ribbon**; the **Customize** dialog box will be displayed. Next, choose the **Keyboard** tab; a list of all the available commands will be displayed, as shown in Figure 1-5. The options corresponding to the **Keyboard** tab are discussed next.

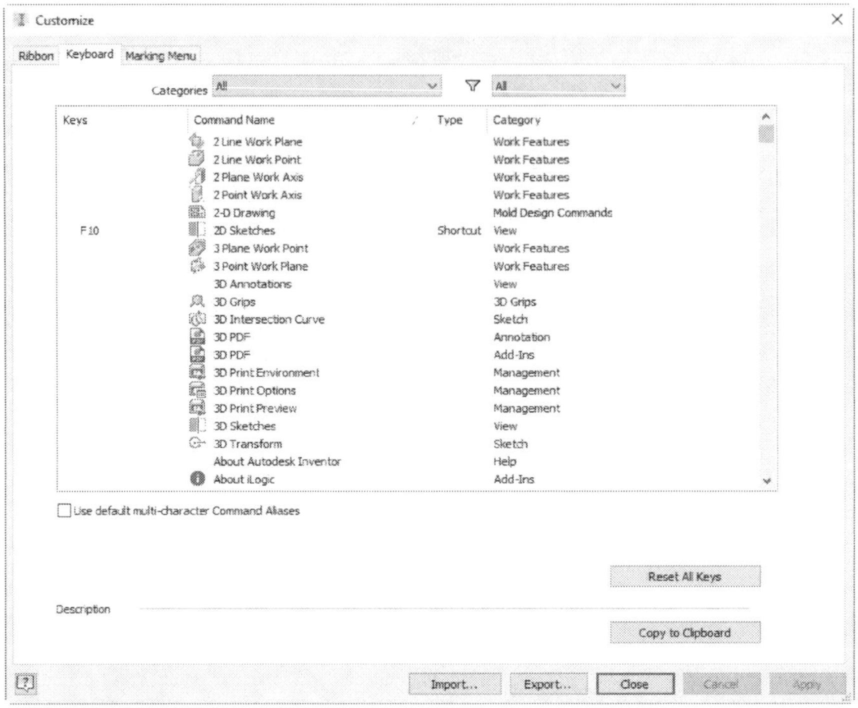

Figure 1-5 *The* ***Customize*** *dialog box displaying various commands in the* ***Keyboard*** *tab*

Categories

Select the required category of command from this drop-down list; the commands related to the selected category will be listed in the list box.

Filter

You can further shortlist the displayed commands from this drop-down list. If you select the **All** option, all the commands related to the selected category will be displayed. If you select the **Assigned** option, then the commands to which the hotkeys are assigned will be displayed. Similarly, if you select the **Unassigned** option, then the commands to which the hotkeys are not assigned will be displayed.

List Box

The list box has four columns: **Keys**, **Command Name**, **Type**, and **Category**. The **Keys** column displays the hotkeys assigned to the commands. The name of the command, its type, and category will be listed in the **Command Name**, **Type**, and **Category** columns, respectively. To assign hotkeys to a tool, click in the **Keys** column that is associated to the command; an edit box will be displayed. In this edit box, enter the shortcut key that you want to assign. To accept the settings, press the Enter key. Else, click on the cross-mark provided next to the tick-mark.

Reset All Keys
The **Reset All Keys** button is used to remove all the customized hotkeys and restore the default hotkeys.

Copy to Clipboard
Choose this button to copy the contents of the **Keyboard** tab and paste them to other document.

Import
Choose this button to restore the customized settings from the .xml format. Note that before importing the file, all the Autodesk Inventor files must be closed.

Export
Choose this button to save the customized settings in the .xml format. Make sure that all the Autodesk Inventor files are closed before choosing this button.

Close
Choose this button to close the **Customize** dialog box.

CREATING THE SKETCH
After starting Autodesk Inventor, you can start creating model in the Part environment. But before creating the model, you need to create its sketch in the Sketching environment. To do so, choose the **Start 2D Sketch** tool from the **Sketch** drop-down in the **Sketch** panel of the **3D Model** tab, see Figure 1-6. On choosing this tool, the Sketching environment is invoked and you can create 2D sketches. If you choose the **Start 3D Sketch** tool from the **Sketch** panel, you can create 3D sketches.

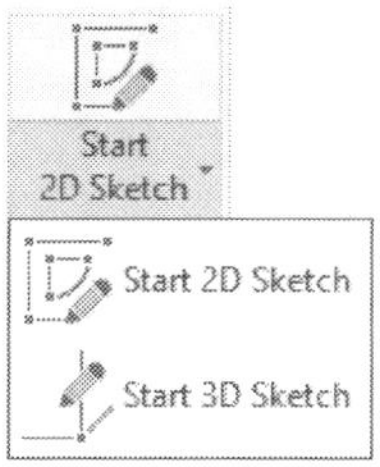

***Figure 1-6** Tools in the **Sketch** drop-down*

MARKING MENU
Marking menu is a type of menu that consists of tools and options which are commonly used in Autodesk Inventor software in different environments. Marking menu replaces the conventional right-click context menu. The Marking menu consists of different tools in different environments. For example, in the Sketching environment, the Marking menu consists of commonly used tools such as **Create Line**, **Two Point Rectangle**, **Done [ESC]**, **Trim**, **General Dimensions,** and so on. In the Modeling environment, it consists of tools and options such as **Extrude**, **Fillet**, **Hole**, **New Sketch**, and so on.

You can invoke a tool in Marking menu by using two modes: Marking mode and Menu mode. To invoke the Marking menu using the Menu mode, right-click anywhere in the graphic window; all the menu items surrounding the cursor will be displayed. After invoking the Marking menu, you can choose the desired tool or option from it. To do so, move the cursor toward the desired tool; the tool is highlighted along with a marker ray. Next, choose the highlighted tool to invoke it.

The other mode, Marking mode, is also known as gesture behavior. It helps you to mark a trail and choose the desired tool. To choose a tool in the Marking mode, right-click and drag the cursor immediately in the direction of the desired tool.

Figure 1-7 shows a Marking menu invoked in the Sketching environment and Figure 1-8 shows a Marking menu which is invoked in the Modeling environment.

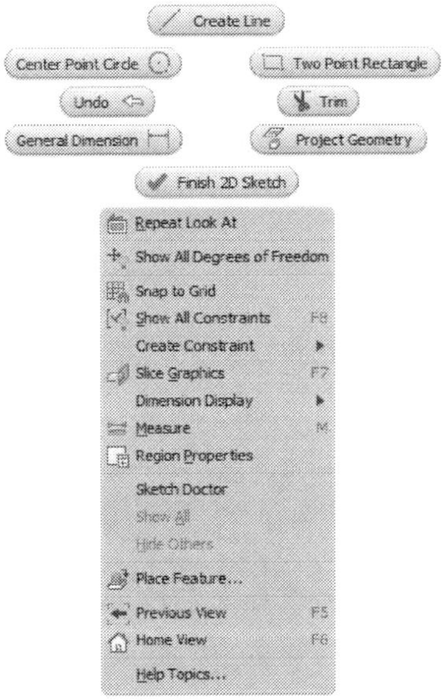

Figure 1-7 *Marking menu available in the Sketching environment*

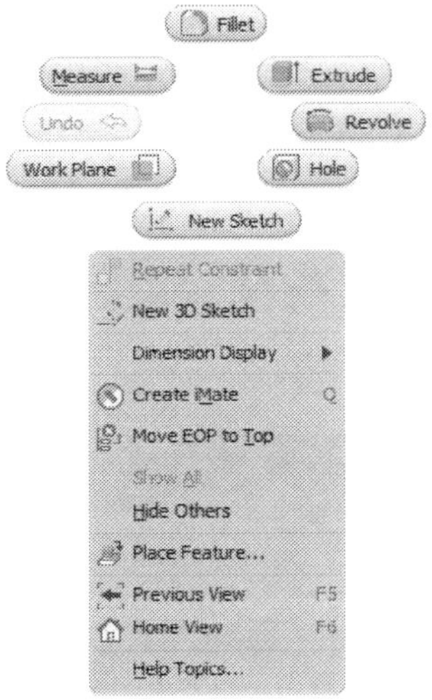

Figure 1-8 *Marking menu available in the Modeling environment*

Tip

You can modify the tools listed in the Marking menu. You can also turn the Marking Menu feature on or off using the options in the ***User Interface*** *flyout in the* ***Windows*** *panel of the* ***View*** *tab in the* ***Ribbon****.*

COLOR SCHEME

Autodesk Inventor allows you to use various color schemes to set the background color of the screen and for displaying the entities on the screen. Note that this book uses the **Presentation** color scheme with a single color background. To change the color scheme, choose the **Application Options** tool from the **Options** panel of the **Tools** tab in the **Ribbon**; the **Application Options** dialog box will be displayed. Choose the **Colors** tab to display the predefined colors. Next, select the **Presentation** option from the **Color scheme** list box in the **Colors** tab. Select **1 Color** from the drop-down list in the **Background** area, refer to Figure 1-9. Choose **Apply** to apply the color scheme to the Autodesk Inventor environment, and then choose **Close**. Note that all the files you open henceforth will use this color scheme.

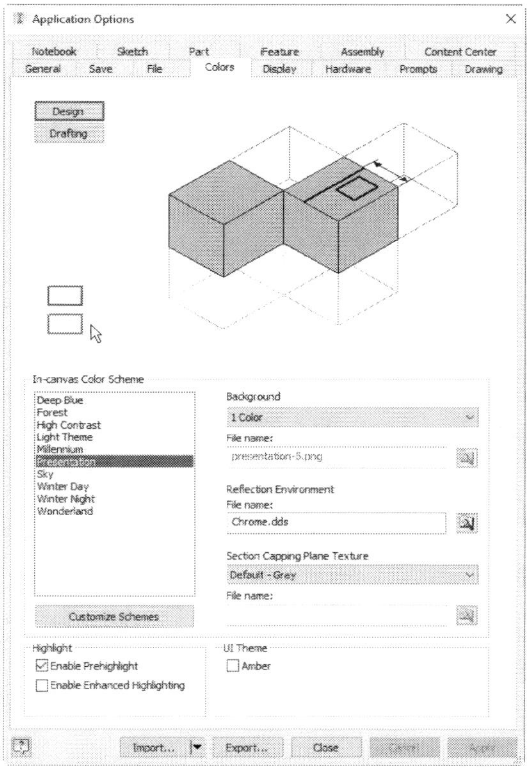

*Figure 1-9 The **Application Options** dialog box with the required options set in the **Colors** tab*

This page is intentionally left blank

Chapter 2

Drawing Sketches for Solid Models

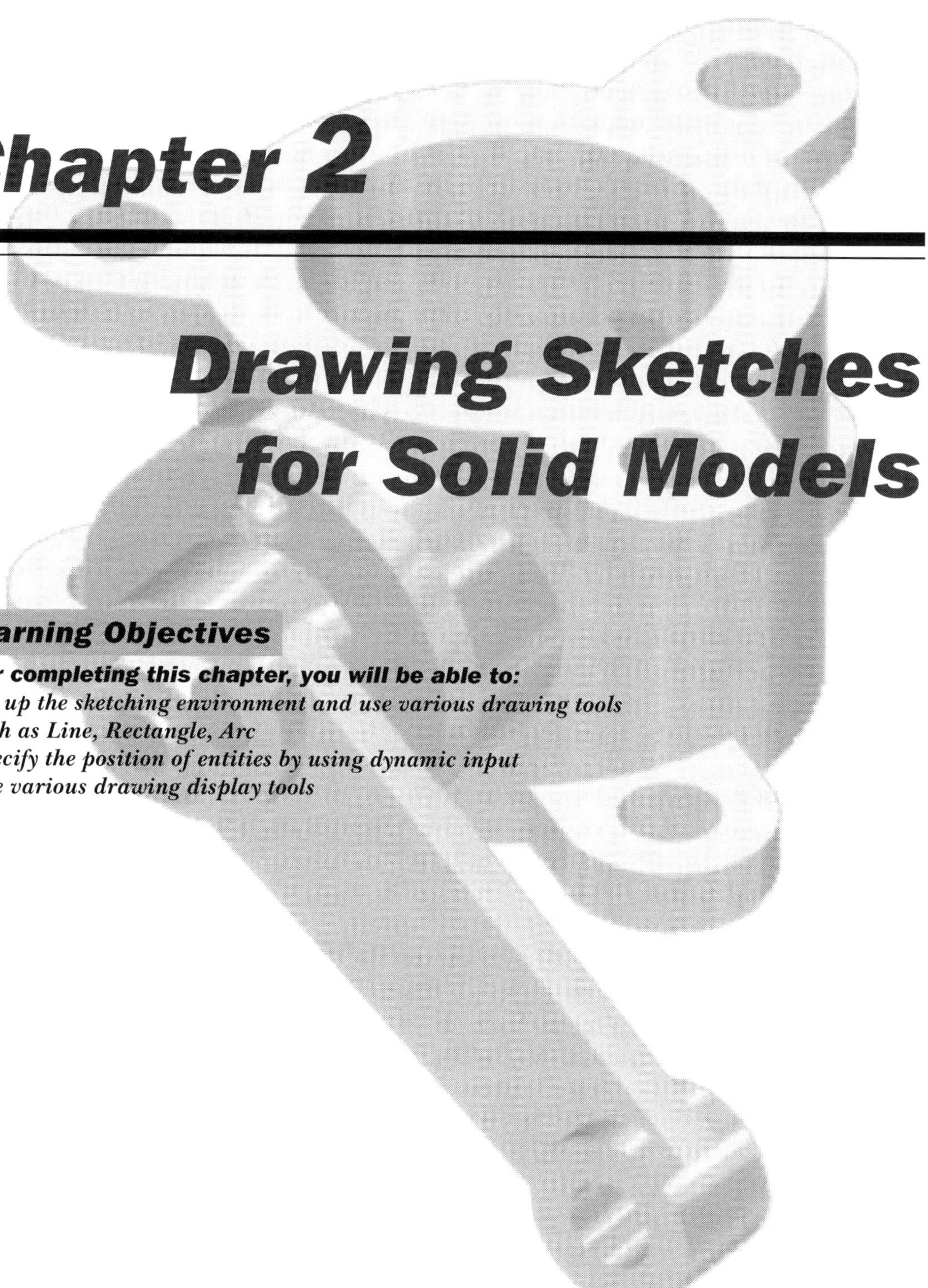

Learning Objectives

After completing this chapter, you will be able to:

- *Set up the sketching environment and use various drawing tools such as Line, Rectangle, Arc*
- *Specify the position of entities by using dynamic input*
- *Use various drawing display tools*

THE SKETCHING ENVIRONMENT

Most of the designs created in Autodesk Inventor consist of sketched and placed features. A sketch is a combination of a number of two-dimensional (2D) entities such as lines, arcs, circles, and so on. The features such as extrude, revolve, and sweep that are created by using 2D sketches are known as sketched features. The features such as fillet, chamfer, thread, and shell that are created without using a sketch are known as placed features. In a design, the base feature or the first feature is always a sketched feature. For example, the sketch shown in Figure 2-1 is used to create the solid model as shown in Figure 2-2. In this figure, the fillets and chamfers are the placed features.

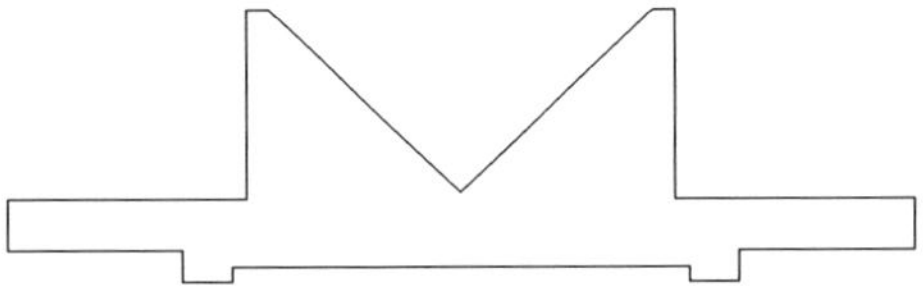

Figure 2-1 *The basic sketch for the solid model*

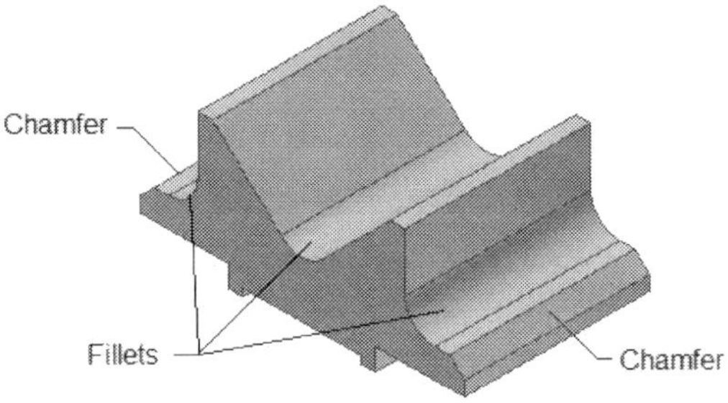

Figure 2-2 *A solid model created using the sketched and 3D model features*

Once you have drawn the basic sketch, refer to Figure 2-1, you need to convert it into a solid model using solid modeling tools.

You can create sketches in the Sketching environment. This environment of Autodesk Inventor can be invoked any time in the **Part** or **Assembly** module. Unlike other solid modeling programs, here you just need to invoke the **Start 2D Sketch** tool and specify the plane to draw sketch, the Sketching environment will be invoked. You can draw a sketch in this environment and then proceed to the part modeling environment for converting the sketch into a solid model. The options in the Sketching environment will be discussed later in this chapter.

Initial Interface of Autodesk Inventor

When you start Autodesk Inventor, the initial interface is displayed with the **Get Started** tab chosen by default, as shown in Figure 2-3. The **Launch** panel of this tab contains options such as **New**, **Open**, **Projects**, and **Open Samples**. By choosing the **Home** option from the **My Home** panel, you can start and open the recent file. The **Team Web** option allows you to attach any required website or HTML file for easy access by entering the website or HTML link in the **Team WEB** area of **File** tab in the **Application Options** dialog box. The **Tutorial Gallery** option displays the guided gallery of Inventor's tutorials. You can navigate to the previous page with the help of the **Back** option. You can also access the Autodesk help and resources from the web by using the **Help** option from the **Help** panel. By choosing the **What's New** option from the **New Features** panel of the **Ribbon**, you can view all the enhancements in Autodesk Inventor 2020. When you choose the **Highlight New** option, the notification badges appear in the ribbon over the options that have enhancements or updates.

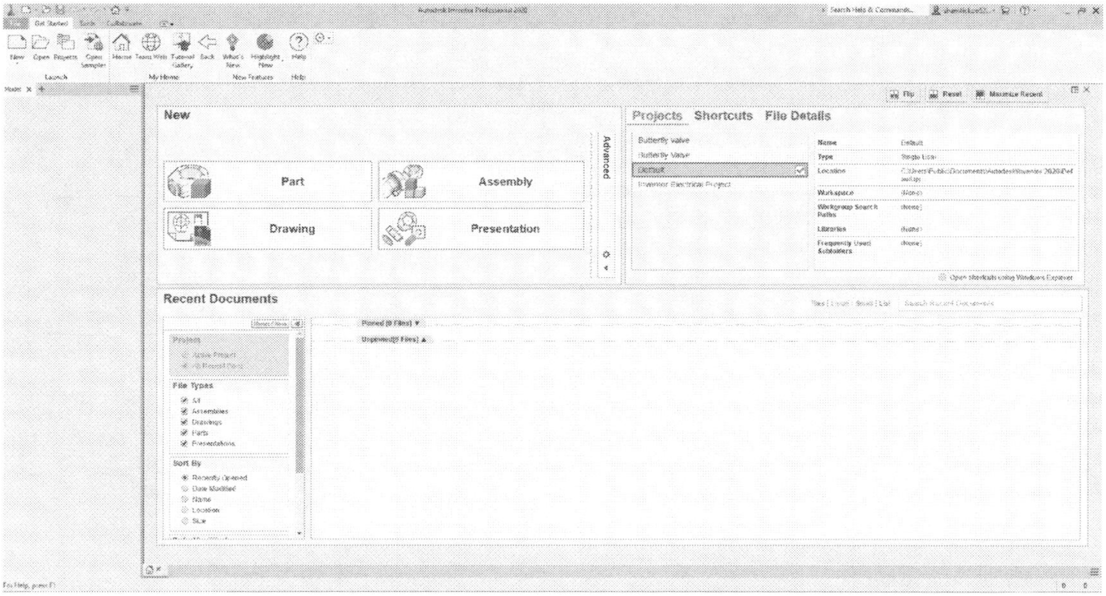

Figure 2-3 The initial interface of Autodesk Inventor Professional 2020

POSITIONING ENTITIES BY USING DYNAMIC INPUT

In Autodesk Inventor, you can specify the position of sketching entities by using the Dynamic Input which consists of two components: Pointer Input and the Dimension Input. The Pointer Input is displayed when you invoke the sketching tools such as **Line**, **Rectangle**, **Arc**, and it displays the coordinates of the current location of the cursor. As you move the cursor, the coordinates change dynamically.

UNDERSTANDING THE DRAWING DISPLAY TOOLS

The drawing display tools or navigation tools are an integral part of any design software. These tools are extensively used during the design process. These tools are available in the **Navigation** Bar located on the right in the graphics window and in the **Navigate** panel of the **View** tab. Some of the drawing display tools in Autodesk Inventor are discussed next.

Zoom All

Ribbon: View > Navigate > Zoom drop-down > Zoom All
Navigation Bar: Zoom flyout > Zoom All

The **Zoom All** tool is used to increase the drawing display area to display all the sketched entities in the current displa

Zoom

Ribbon:	View > Navigate > Zoom drop-down > Zoom
Navigation Bar:	Zoom flyout > Zoom

The **Zoom** tool is used to interactively zoom in and out of the drawing view. When you choose this tool, the default cursor is replaced by a zoom cursor. You can zoom in the drawing by pressing the left mouse button and dragging the cursor down. Similarly, you can zoom out the drawing by pressing the left mouse button and then dragging the cursor up.

Zoom Window

Ribbon:	View > Navigate > Zoom drop-down > Zoom Window
Navigation Bar:	Zoom flyout > Zoom Window

The **Zoom Window** tool is used to define an area to be magnified and viewed in the current drawing.

Zoom Selected

Ribbon:	View > Navigate > Zoom drop-down > Zoom Selected
Navigation Bar:	Zoom flyout > Zoom Selected

When you choose the **Zoom Selected** tool, you will be prompted to select an entity to zoom. Select an entity from the drawing area; it will be magnified to the maximum extent and will be placed at the center of the drawing window.

Pan

Ribbon:	View > Navigate > Pan
Navigation Bar:	Pan

The **Pan** tool is used to drag the current view in the drawing window. This option is generally used to display the contents of the drawing that are outside the display area without actually changing the magnification of the current drawing.

Orbit

Ribbon:	View > Navigate > Orbit drop-down > Orbit
Navigation Bar:	Orbit flyout > Orbit

The **Orbit** tool is used to rotate a model freely about any axis. It is useful when you want to rotate a model to any position.

Constrained Orbit

Ribbon:	View > Navigate > Orbit drop-down > Constrained Orbit
Navigation Bar:	Orbit flyout > Constrained Orbit

The **Constrained Orbit** tool is used to visually maneuver around the 3D objects to obtain different views.

TUTORIALS

Tutorial 1

In this tutorial, you will draw the sketch of the model shown in Figure 2-4. The sketch to be drawn is shown in Figure 2-5. Do not dimension it, as the dimensions are given only for reference. **(Expected time: 30 min)**

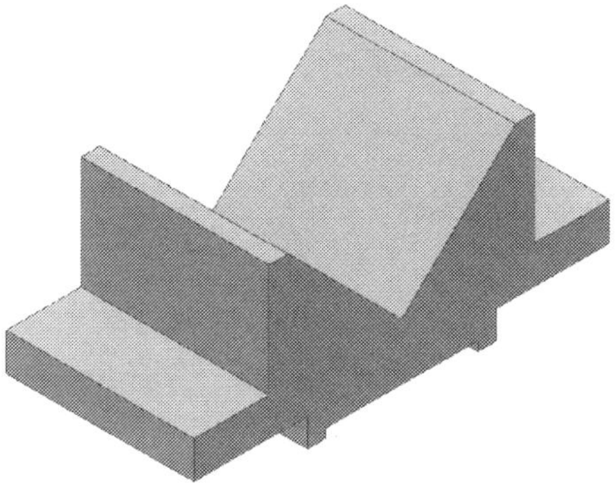

Figure 2-4 *Model for Tutorial 1*

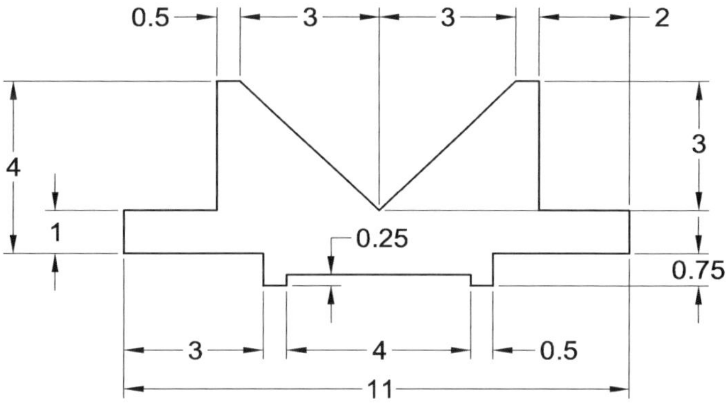

Figure 2-5 *Sketch of the model*

The following steps are required to complete this tutorial:

a. Start a new Autodesk Inventor session and then start a new metric part file.
b. Invoke the Sketching environment and draw the sketch by specifying the coordinates of the points in the Dynamic Input by using the **Line** tool.
c. Save the sketch with the name *Tutorial1* and close the file.

Starting Autodesk Inventor

1. Start Autodesk Inventor by double-clicking on its shortcut icon on the desktop of your computer; a new session of Autodesk Inventor is started.

2. Choose the **New** tool from the **Launch** panel of the **Ribbon**; the **Create New File** dialog box is displayed.

3. Choose the **Metric** template and then double-click on the **Standard (mm).ipt** icon to start a standard metric template; a new metric standard part file starts.

4. Choose the **Start 2D Sketch** tool from the **Sketch** panel of the **3D Model** tab; the default planes are displayed and you are prompted to select the sketching plane.

5. Now, select the **YZ** plane as the sketching plane from the graphics window; the Sketching environment is invoked and the **YZ** plane becomes parallel to the screen.

Note

1. If, by default, the Grid lines are not displayed in the sketching environment, choose the ***Application Options*** *tool from the* ***Options*** *panel of the* ***Tools*** *tab; the* ***Application Options*** *dialog box will be invoked. Now, select the* ***Grid lines*** *check box from the* ***Display*** *area of the* ***Sketch*** *tab.*

2. For the purpose of accuracy, grid lines are turned on in all the tutorials.

Drawing the Sketch

As mentioned earlier, Autodesk Inventor is parametric in nature. Therefore, you can draw the sketch from any point in the drawing window. In this tutorial, Dynamic Input has been used to enter dimensions while drawing the sketch.

1. Choose the **Line** tool from the **Create** panel in the **Sketch** tab. Alternatively, choose the **Create Line** tool from the Marking Menu that is displayed on right-clicking in the drawing / graphics window. On doing so, you are prompted to select the first point for the line. Next, choose **Precise Input** from the expanded **Create** panel of the **Sketch** tab; the **Inventor Precise Input** toolbar is displayed. Double-click on the title bar of this toolbar to dock it. If required, you can also leave this toolbar floating on the screen.

   Initially, the **Precise Input** button is not enabled. This button is enabled only when you invoke a sketching tool. Since all initial settings are configured, you can now start drawing the sketch.

   When you invoke the **Line** tool, the cursor is replaced by a drawing cursor that has a yellow circle at the intersection of crosshairs. When you move the crosshair in the drawing window, this circle snaps to the point that is closer to it. Also, coordinates of the current location of the cursor are displayed at the status bar.

2. To specify the first point, enter **0** in both the **X** and **Y** edit boxes of the **Inventor Precise Input** toolbar and then press ENTER; you are prompted to specify the endpoint of the line.

Tip

You can use the TAB key to switch from the ***X*** *edit box to the* ***Y*** *edit box and vice versa in the* ***Inventor Precise Input*** *toolbar.*

3. In the **Inventor Precise Input** toolbar, enter **3** and **3** in the **X** and **Y** edit boxes, respectively. Next, press ENTER to define the endpoint of the line. On doing so, the first line of the sketch is drawn and you are prompted to select the end of the line.

   You will notice that the dimensions of the sketch are very small but the drawing display area is large. Therefore, you need to modify the drawing display area by using the drawing display tools. To do so, you can use the **Zoom** tool.

4. Choose the **Zoom** tool from the **Navigation Bar**; the drawing cursor is changed to an arrow cursor.

5. Move the cursor to the top of the drawing window, press and hold the left mouse button and then drag the cursor downward till the display is adjusted.

6. Right-click to display the shortcut menu, and then choose **Done** to exit the **Zoom** tool.

7. Again, right-click in the graphics window and then choose **Cancel (ESC)** from the Marking Menu to exit the **Line** tool. Also, close the **Inventor Precise Input** toolbar.

8. Invoke the **Line** tool again. From this step onward, you will use the Dynamic Input to create lines in this tutorial. Select the end point of the first line that was created with the help of the **Inventor Precise Input** toolbar. Move the cursor toward left, and enter **0.5** in the length input field and **90** in the angle input field and then press ENTER. Use the TAB key to toggle between the input fields.

9. Now, you need to create the third line between points 3 and 4, refer to Figure 2-6. To do so, move the cursor downward in the graphics window. Enter **3** in the length input field, press TAB to switch to the angle input field, and enter **180** and then press ENTER; a line is created between the 3 and 4 points.

Note

*The angular and linear dimensions have not been shown in Figure 2-6 for clarity of sketches. You can control the display of the linear and angular dimensions of the sketch. To do so, choose **Application Options** from the **Options** panel of the **Tools** tab; the **Application Options** dialog box is displayed. In the **Sketch** tab of the **Application Options** dialog box, choose the **Settings** button from the **Constraint Settings** area; the **Constraint Settings** dialog box is displayed. Clear the **Create dimensions from input values** check box in the **Dimension** area of the **Constraint Settings** dialog box and then choose **OK**. Next, choose the **Apply** and **Close** buttons from the **Application Options** dialog box to turn off applying dimensions automatically while sketching.*

10. Move the cursor toward left in the graphics window. Enter **2** in the length input field, press TAB to switch to the angle input field, and enter **90** in it, and then press ENTER; a line is created between points 4 and 5, refer to Figure 2-6.

11. Move the cursor downward in the graphics window. Enter **1** in the length input field, press TAB, and enter **90** in the angle input field and then press ENTER; a line is created between points 5 and 6, refer to Figure 2-6.

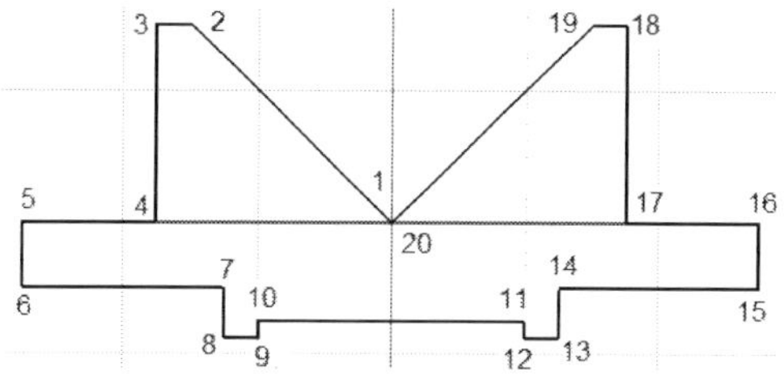

Figure 2-6 *Sketch for Tutorial 1*

12. Move the cursor toward right in the graphics window. Enter **3** in the length input field, press TAB, and enter **90** in the angle input field and then press ENTER; a line is created between points 6 and 7.

13. Move the cursor downward in the graphics window. Enter **0.75** in the length input field, press TAB, and enter **90** in the angle input field and then press ENTER; a line is created between points 7 and 8, refer to Figure 2-6.

14. Move the cursor toward right in the graphics window. Enter **0.5** in the length input field, press TAB, and enter **90** in the angle input field and then press ENTER; a line is created between points 8 and 9, refer to Figure 2-6.

15. Move the cursor upward in the graphics window. Enter **0.25** in the length input field, press TAB, and enter **90** in the angle input field and then press ENTER; a line is created between points 9 and 10, refer to Figure 2-6.

16. Move the cursor toward right in the graphics window. Enter **4** in the length input field, press TAB, and enter **90** in the angle input field and then press ENTER; a line is created between points 10 and 11, refer to Figure 2-6.

17. Move the cursor downward in the graphics window. Enter **0.25** in the length input field, press TAB, and enter **90** in the angle input field and then press ENTER; a line is created between points 11 and 12, refer to Figure 2-6.

18. Move the cursor toward right in the graphics window. Enter **0.5** in the length input field, press TAB and enter **90** in the angle input field and then press ENTER; a line connecting points 12 and 13 is created, refer to Figure 2-6.

19. Move the cursor upward in the graphics window. Enter **0.75** in the length input field, press TAB, and enter **90** in the angle input field and then press ENTER; a line connecting points 13 and 14 is created, refer to Figure 2-6.

20. Move the cursor toward right in the graphics window. Enter **3** in the length input field and **90** in the angle input field and then press ENTER; a line connecting points 14 and 15 is created, refer to Figure 2-6.

21. Move the cursor upward in the graphics window. Enter **1** in the length input field, press TAB, and enter **90** in the angle input field and then press ENTER; a line connecting points 15 and 16 is created, refer to Figure 2-6.

22. Move the cursor toward left in the graphics window. Enter **2** in the length input field, press TAB, and enter **90** in the angle input field and then press ENTER; a line connecting points 16 and 17 is created, refer to Figure 2-6.

23. Move the cursor upward in the graphics window. Enter **3** in the length input field, press TAB, and enter **90** in the angle input field and then press ENTER; a line connecting points 17 and 18 is created, refer to Figure 2-6.

24. Move the cursor toward left in the graphics window. Enter **0.5** in the length input field, press TAB, and enter **90** in the angle input field and then press ENTER; a line connecting points 18 and 19 is created, refer to Figure 2-6.

25. Now, to close the geometry, click on the point 1 that you created with the help of the **Inventor Precise Input** toolbar.

26. Next, right-click in the graphics window and choose **Cancel (ESC)** from the Marking menu displayed to exit the **Line** tool.

Saving the Sketch

Remember that you cannot save a sketch in the Sketching environment. This is because the Sketching environment is just a part of the **Part** module in Autodesk Inventor. This environment is used only for drawing the sketches of features. Therefore, you need to exit the Sketching environment to save the sketch for further use. The sketches in the **Part** module are saved in the *.ipt* format.

1. Right-click in the graphics window and then choose the **Finish 2D Sketch** button from the Marking Menu displayed; the Sketching environment is closed and you switch to the Part modeling environment. Now, choose the **Home** button of the ViewCube; the current orientation of the sketch is changed to Isometric. Also, notice that the **3D Model** tab is activated in place of the **Sketch** tab. The options in the **3D Model** tab are used to create features. The options under this tab will be discussed in later chapters.

2. Choose **Save** from **Quick Access Toolbar**; the **Save As** dialog box is displayed, as shown in Figure 2-7.

3. Create a new folder with the name *Inventor_2020* in the C drive of your computer. In this folder, create another folder with the name *c02*.

4. Enter **Tutorial1** as the file name in the **File name** edit box, refer to Figure 2-7, and then choose the **Save** button from the **Save As** dialog box to save the sketch.

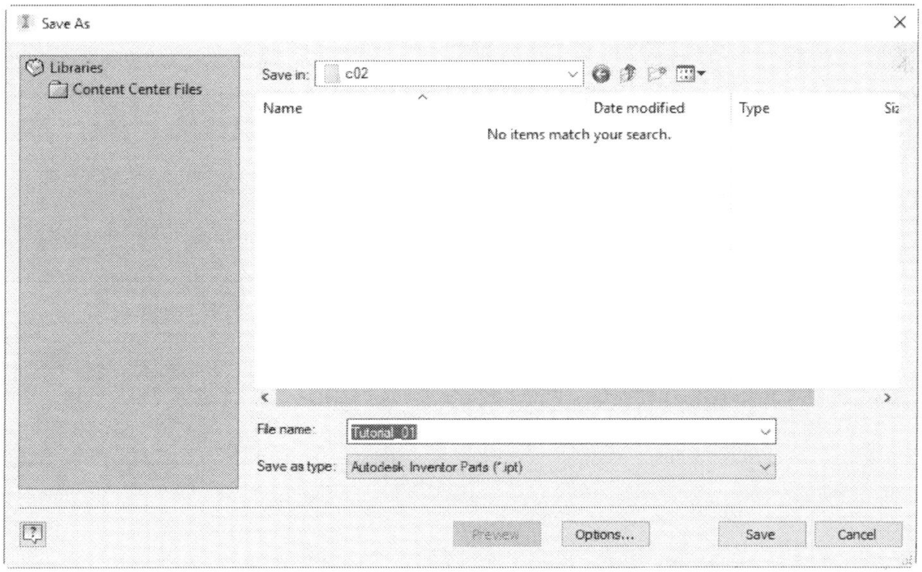

Figure 2-7 *The* ***Save As*** *dialog box*

5. Choose **Close > Close** from the **File Menu** to close this file.

Tutorial 2

In this tutorial, you will draw the basic sketch of the revolved solid model shown in Figure 2-8. The sketch for creating this model is shown in Figure 2-9. Do not dimension the sketch as the dimensions are given only for reference. Use Dynamic Input to draw the feature.

(Expected time: 30 min)

The following steps are required to complete this tutorial:

a. Start a new metric standard part file and invoke the Sketching environment.
b. Draw the sketch with the help of Dynamic Input by using help of **Line** tool.
c. Draw fillets.
d. Save the sketch with the name Tutorial2 and close the file.

Figure 2-8 *Revolved model for Tutorial 2*

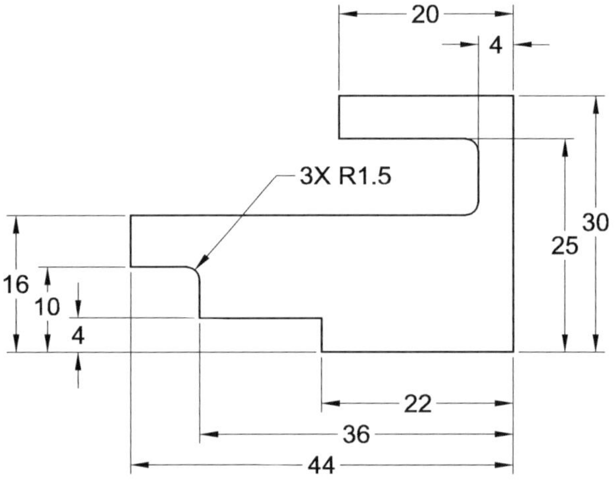

Figure 2-9 *Sketch for the revolved model*

Starting a New File

1. Choose the **New** tool from the **Launch** panel of the **Get Started** tab to invoke the **Create New File** dialog box.

2. Choose **Metric** to display the standard metric templates. Double-click on **Standard (mm).ipt** to start a new metric part file.

3. Choose the **Start 2D Sketch** tool from the **Sketch** panel of the **3D Model** tab; the default planes are displayed and you are prompted to select the sketching plane.

4. Now, select the **YZ** plane as the sketching plane from the graphics window; the Sketching environment is invoked and the **YZ** Plane becomes parallel to the screen.

Drawing the Sketch

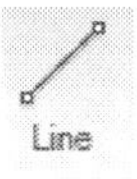

1. Choose the **Line** tool from the **Create** panel in the **Sketch** tab or from the Marking Menu. On doing so, you are prompted to select the first point of the line to be created. In the Dynamic Input, press TAB and enter **0** in the X co-ordinate field. Next, press TAB, and enter **0** in the Y coordinate field and press ENTER.

2. Now, you need to draw the line 1. Move the cursor toward left, enter **22** in the length input field, and enter **90** in the angle input field of the Dynamic Input and press ENTER; the line 1 is drawn.

3. Move the cursor upward in the graphics window. Next, enter **4** in the length input field, press TAB, and enter **90** in the angle input field and press ENTER; the line 2 is drawn, as shown in Figure 2-10.

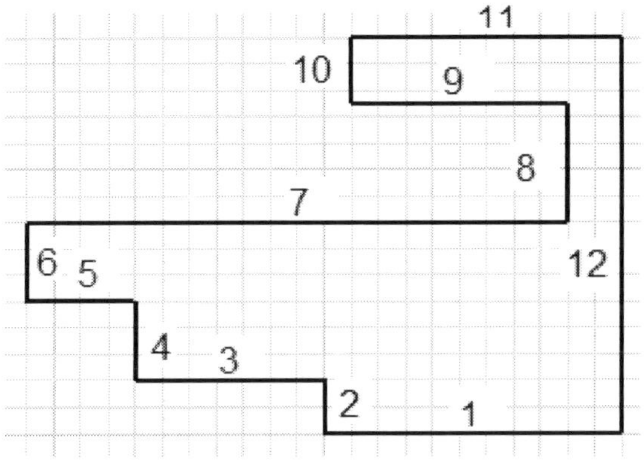

Figure 2-10 *Sketch after drawing the lines*

4. Move the cursor toward left in the graphics window. Next, enter **14** in the length input field, press TAB, and enter **90** in the angle input field in the Dynamic Input and press ENTER; line 3 is drawn.

5. Move the cursor upward in the graphics window. Next, enter **6** in the length input field, press TAB, and enter **90** in the angle input field in the Dynamic Input and press ENTER; line 4 is drawn.

6. Move the cursor toward the left in the graphics window. Next, enter **8** in the length input field, press TAB, and enter **90** in the angle input field in the Dynamic Input and press ENTER; line 5 is drawn.

7. Move the cursor upward in the graphics window. Next, enter **6** in the length input field, press TAB, and enter **90** in the angle input field in the Dynamic Input and press ENTER; line 6 is drawn.

8. Move the cursor toward the right in the graphics window. Next, enter **40** in the length input field, press TAB, and enter **90** in the angle input field and press ENTER; line 7 is drawn.

9. Move the cursor upward in the graphics window. Next, enter **9** in the length input field, press TAB, and enter **90** in the angle input field and press ENTER; line 8 is drawn.

10. Move the cursor toward left in the graphics window. Next, enter **16** in the length input field, press TAB, and enter **90** in the angle input field and press ENTER; line 9 is drawn.

11. Move the cursor upward in the graphics window. Next, enter **5** in the length input field, press TAB, and then enter **90** in the angle input field and press ENTER; line 10 is drawn.

12. Move the cursor toward the right in the graphics window. Next, enter **20** in the length input field, press TAB, and enter **90** in the angle input field and press ENTER; line 11 is drawn.

13. Move the cursor downward in the graphics window. Next, enter **30** in the length input field and **90** in the angle input field and press ENTER; line 12 is drawn.

14. The initial sketch is drawn. Exit the **Line** tool by choosing **OK** from the Marking menu.

Drawing Fillets

1. Choose the **Fillet** tool from the **Sketch > Create > Fillet/Chamfer** drop-down; the **2D Fillet** dialog box is displayed. Enter **1.5** in the **Radius** edit box of this dialog box. Do not press ENTER.

2. Select the line 8 and then line 9, refer to Figure 2-10; a fillet is created between these lines and the radius of the fillet is displayed in the sketch.

3. Similarly, select lines 7 and 8 and then lines 4 and 5 to create a fillet between these lines. Next, right-click, and choose **OK** from the Marking Menu to exit the **Fillet** tool after creating all fillets.

   As all the lines are filleted with the same radius value, the radius of the fillet is not displayed on other fillets. This completes the sketch. The final sketch for this tutorial after filleting all the sketches is shown in Figure 2-11.

Note

In Figure 2-11, the display of the dimensions, axes and grids have been turned off for a better visibility of the lines of the sketch.

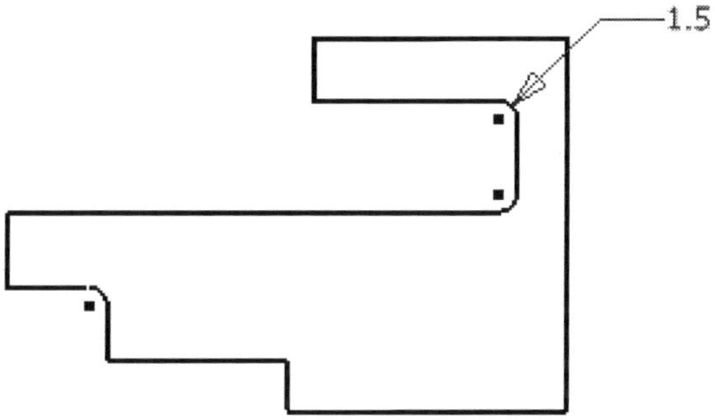

Figure 2-11 *Final sketch after filleting*

Saving the Sketch

1. Choose the **Finish 2D Sketch** button from the Marking Menu.

2. Choose the **Save** button from **Quick Access Toolbar** and save this sketch with the name *Tutorial4* at the location given below.
C:\Inventor_2020\c02

3. Choose **Close > Close** from the File Menu to close this file.

EXERCISES

Exercise 1

Draw the basic sketch of the model shown in Figure 2-12. The sketch to be drawn is shown in Figure 2-13. Do not dimension it as the dimensions are given only for reference.

(Expected time: 30 min)

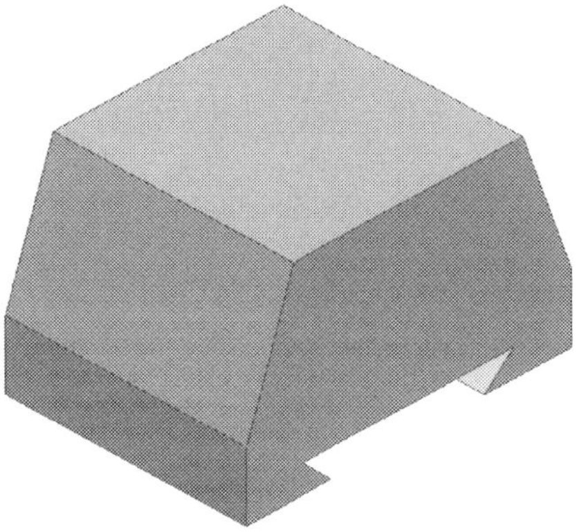

Figure 2-12 Model for Exercise 1

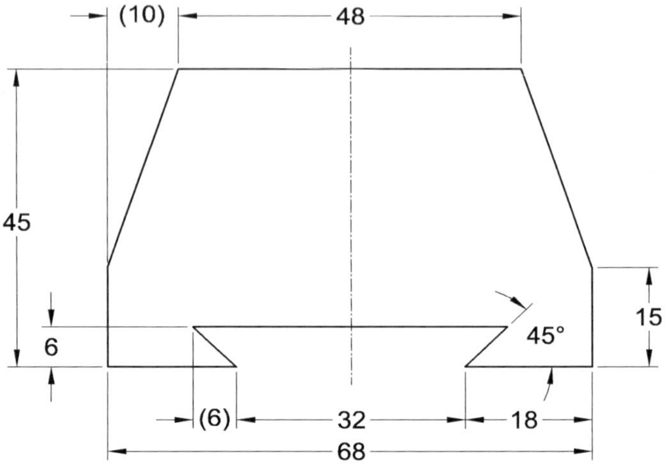

Figure 2-13 Sketch for Exercise 1

Exercise 2

Draw the basic sketch of the model shown in Figure 2-14. The sketch to be drawn is shown in Figure 2-15. Do not dimension it as the dimensions are given only for reference.

(Expected time: 45 min)

Figure 2-14 Model for Exercise 2

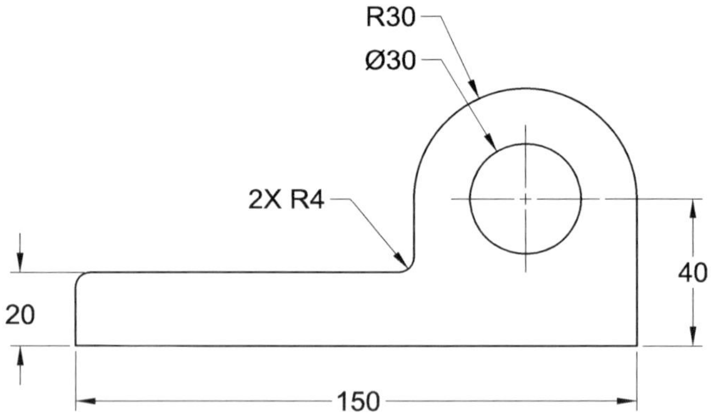

Figure 2-15 Sketch for Exercise 2

Chapter 3

Adding Constraints and Dimensions to Sketches

Learning Objectives

After completing this chapter, you will be able to:

- *Add geometric constraints to a sketch*
- *Control the constraint inference*
- *View and delete constraints from a sketch*
- *Dimension a sketch*
- *Modify the dimensions of a sketch*
- *Measure distances, angles, loops, and areas in a sketch*

ADDING GEOMETRIC CONSTRAINTS TO A SKETCH

Constraints are applied to the sketched entities to define their size and position with respect to other elements. Also, they are useful for capturing the design intent. As mentioned in Chapter 1, there are twelve types of geometric constraints that can be applied to the sketched entities. These constraints restrict their degrees of freedom and make them stable. Most of these constraints get automatically applied to the entities while drawing. However, sometimes you may need to apply some additional constraints to the sketched entities. These constraints are discussed next.

Perpendicular Constraint

Ribbon:	Sketch > Constrain > Perpendicular Constraint

The Perpendicular constraint forces the selected entity to become perpendicular to the specified entity. The entities to which the constraints can be applied are lines and ellipse axes.

Parallel Constraint

Ribbon:	Sketch > Constrain > Parallel Constraint

The Parallel constraint forces the selected entity to become parallel to the specified entity. The entities to which this constraint can be applied are lines and ellipse axes.

Tangent Constraint

Ribbon:	Sketch > Constrain > Tangent

The Tangent constraint forces the selected line segment or curve to become tangent to another curve.

Coincident Constraint

Ribbon:	Sketch > Constrain > Coincident Constraint

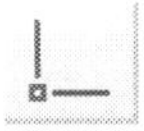

The Coincident constraint forces two points or a point and a curve to become coincident.

Concentric Constraint

Ribbon:	Sketch > Constrain > Concentric Constraint

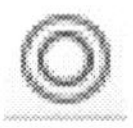

The Concentric constraint forces two curves to share the same location of center points. The curves that can be made concentric include arcs, circles, and ellipses.

Collinear Constraint

Ribbon:	Sketch > Constrain > Collinear Constraint

The Collinear constraint forces the selected line segments or ellipse axes to be placed in the same line.

Horizontal Constraint

Ribbon:	Sketch > Constrain > Horizontal Constraint

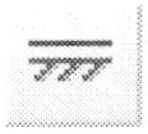

The Horizontal Constraint forces the selected line segment, ellipse axis, or two points to become horizontal, irrespective of their original orientation.

Vertical Constraint

Ribbon:	Sketch > Constrain > Vertical Constraint

The Vertical constraint is similar to the Horizontal constraint with the only difference that this constraint forces the selected entities to become vertical.

Equal Constraint

Ribbon:	Sketch > Constrain > Equal

The Equal constraint can be used for line segments or curves. If you select two line segments, this constraint will force the length of one of the selected line segments to become equal to the length of the other selected line segment.

Fix Constraint

Ribbon:	Sketch > Constrain > Fix

The Fix constraint is used to fix the orientation or location of the selected curve or point with respect to the coordinate system of the current drawing.

Symmetric Constraint

Ribbon:	Sketch > Constrain > Symmetric

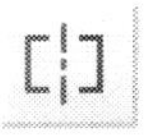

This constraint is used to force two selected sketched entities to become symmetrical about a single sketched line segment.

Smooth Constraint

Ribbon:	Sketch > Constrain > Smooth (G2)

This constraint is used to apply curvature continuity between a spline and an entity connected to it. The entities that can be selected to apply this constraint include a line, arc, or another spline. Note that these entities should be connected to the spline.

CONTROLLING CONSTRAINTS AND APPLYING THEM AUTOMATICALLY WHILE SKETCHING

Ribbon: Sketch > Constrain > Constraint Settings

You can control and select the constraints that need to be applied automatically as well as select the geometry to which they will be applied. You can do so by using the **Constraint Settings** dialog box.

DELETING GEOMETRIC CONSTRAINTS

Autodesk Inventor allows you to delete the constraints applied to the selected entities. To delete constraints, first you need to show the constraint of entities by using the **Show Constraints** tool. Once the constraints are displayed, exit the **Show Constraints** tool by pressing the ESC key. Next, move the cursor over the constraint that you want to delete; it will be highlighted in red. Click the left mouse button to select the constraint. Next, move the cursor away and right-click, and then choose **Delete** from the Marking menu.

MEASURING SKETCHED ENTITIES

Ribbon: Inspect > Measure > Measure

Autodesk Inventor allows you to measure various parameters of the sketched entities. The parameters that you can measure are distances, angles, loops, and area. All these parameters can be measured by using the **Measure** tool from the **Measure** panel of the **Inspect** tab. The various methods of measuring these parameters are discussed next.

TUTORIALS

Tutorial 1

In this tutorial, you will draw the sketch for the model shown in Figure 3-1. After drawing the sketch, you will add the required constraints to it and then dimension it. The dimensioned sketch required for this model is shown in Figure 3-2. The solid model shown in Figure 3-1 is only for reference.

The sketch shown in Figure 3-2 is the combination of multiple closed loops: the outer loop and inner circles. As the numbers of loops increase, so does the complexity of the sketch. This is because the numbers of constraints and dimensions in a sketch increase in case of multiple loops. Now, to draw sketches without using the Dynamic Input, it is recommended that you first draw the outer loop of the sketch and then add constraints and dimensions to it. This is because once the outer loop has been constrained and dimensioned, the inner circles can be constrained and dimensioned easily with reference to the outer loop.

(Expected time: 30 min)

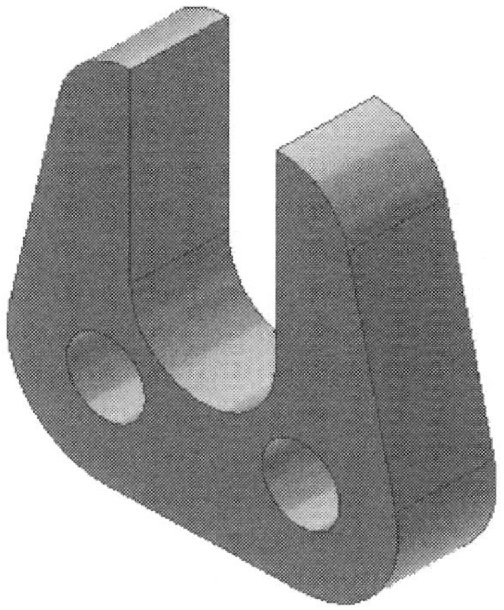

Figure 3-1 Model for Tutorial 3

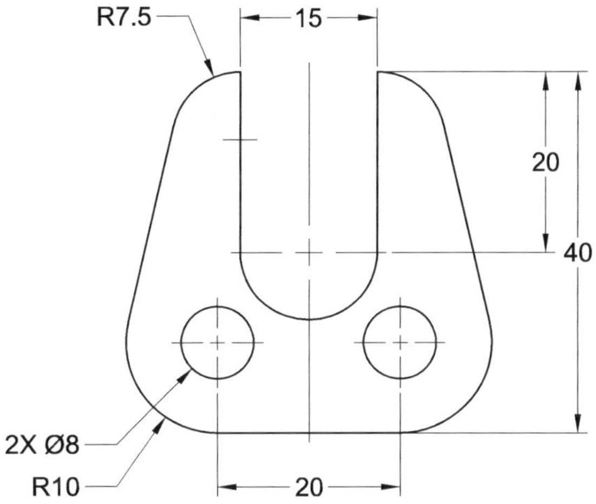

Figure 3-2 Dimensioned sketch of the model

The following steps are required to complete this tutorial:

a. Start a new metric template and draw the outer loop of the sketch.
b. Add the required dimensions and constraints to the outer loop.
c. Draw inner circles and add constraints and dimensions to them.
d. Save the sketch with the name *Tutorial1.ipt* and close the file.

Starting a New File and Invoking the Sketching Environment

1. Choose the **New** button from the **Quick Access Toolbar** and start a new metric standard part file using the **Metric** tab of the **Create New File** dialog box.

2. Choose the **Start 2D Sketch** button from the **Sketch** panel of the **3D Model** tab; the default planes are displayed and you are prompted to select the sketching plane.

3. Select the **XY Plane** as the sketching plane from the graphics window; the Sketching environment is invoked and the **XY Plane** becomes parallel to the screen.

Drawing the Outer Loop

1. Invoke the **Line** tool from the **Create** panel in the **Sketch** tab, refer to Figure 3-3.

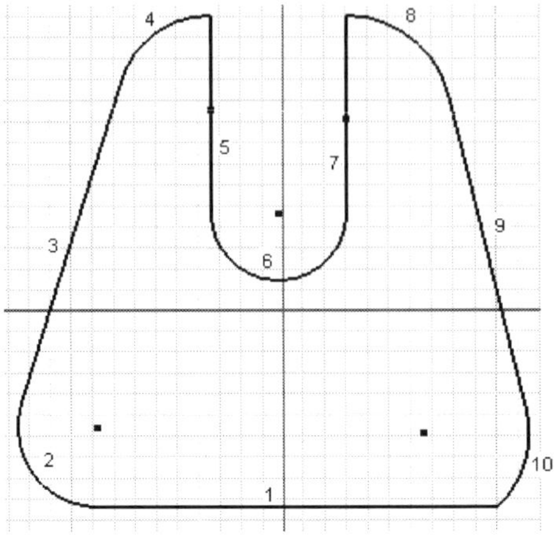

Figure 3-3 *Profile with geometries numbered*

You can also draw the tangent arcs while using the **Line** tool. This can be done by invoking the **Line** tool and by pressing the left mouse button and then dragging it in the required direction. Move the cursor close to the endpoint of the last line until the yellow circle snaps to that point. When the yellow circle snaps to the endpoint, it turns gray. Next, press and hold the left mouse button and drag the mouse through a small distance in the upward direction.

For your reference, all the geometries in the sketch are numbered. You will draw inner holes in the sketch after dimensioning the outer loop.

Note

The outer loop that you created in the previous step might be different from the one shown in Figure 3-3. You can find the missing constraints by following the next step and applying constraints accordingly.

Adding Constraints to Sketched Entities

As evident from Figure 3-3, some of the constraints such as tangent and equal are missing in the sketch. The sketch shown in Figure 3-3 may not be symmetrical and all the lines in the sketch may not be tangent to the arcs. Therefore, you need to add these missing constraints manually to the sketch to complete it. You can choose the **Show Constraint** option from the Marking menu that is displayed when you right-click on an entity.

1. In Figure 3-3, the **Tangent** constraint is missing between line 1 and arc 10. To add this constraint between the line and the arc, choose the **Tangent** tool from the **Constrain** panel of the **Sketch** tab; you are prompted to select the first curve. Select arc 10 as the first curve; you are prompted to select the second curve. Select line 1 as the second curve. Similarly, add this constraint to all the places in the sketch wherever it is missing.

The geometries 5 and 7, and 3 and 9 are the lines that must be of equal length. Also, the geometries 2 and 10, and 4 and 8 are the arcs that must be of equal radii. Therefore, you need to add the Equal constraint between the respective pairs of all these geometries.

2. Choose the **Equal** tool from the **Constrain** panel of the **Sketch** tab.

3. Select line 5 as the first line and then line 7 as the second line to apply the Equal constraint to them; you are prompted again to select the first entity.

4. Select line 3 and then line 9 to apply the **Equal** constraint to them; you are prompted to select the first entity again.

5. Select arc 2 and then arc 10 to apply the **Equal** constraint to them. Applying this constraint to arcs or circles forces their radii or diameters to be equal.

6. Similarly, apply the **Equal** constraint to arc 4 and arc 8.

7. Apply the **Coincident Constraint** between the center point of arc 4 and line 5, and the center point of arc 8 and line 7 if not automatically applied.

8. Choose the **Symmetric** tool from the **Constrain** panel of the **Sketch** tab to apply **Symmetric** constraint between line 3 and line 9. To apply a symmetric constraint you need to draw a vertical line of symmetry through the origin.

9. Choose the **Coincident Constraint** tool from the **Constrain** panel; you are prompted to select the first curve or point.

10. Select the origin; you are prompted to select the second curve or point.

11. Select the center point of arc 6; the entire sketch moves to make the origin coincident with the center point of the arc. The sketch after applying all the constraints is shown in Figure 3-4.

Note

The shape of the sketch that you have drawn may be a little different from the final sketch at this stage because of the difference in specifying points while drawing the sketch. However, once all the dimensions are applied, the shape of the sketch will be the same as of the final sketch. Also, you may need to add vertical constraint to lines 5 and 7 and horizontal constraint to line 1 to fully constrain the sketch.

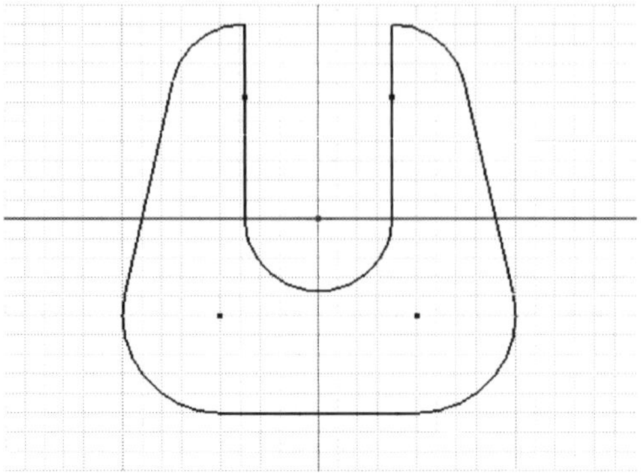

Figure 3-4 *The sketch after applying all constraints*

Dimensioning the Sketch

1. Choose the **Dimension** tool from the **Constrain** panel of the **Sketch** tab. Next, right-click to display the Marking menu. From the Marking menu, choose **Edit Dimension** if the check mark is not available on the left of the **Edit Dimension** option. If it shows the check mark, press the ESC key once to exit the Marking Menu. On doing so, you are prompted to select the geometry to be dimensioned. Select line 1 and place the dimension below the sketch. Modify the value of this dimension in the **Edit Dimension** edit box to **20**.

2. Select arc 4 and place the dimension on the left of the sketch; the radius dimension of the sketch is placed. Modify the dimension value in the **Edit Dimension** edit box to **7.5**. The size of arc 8 is also modified because the Equal constraint is applied between these two entities.

Note

As discussed in the previous tutorial, you may need to use the combination of hot keys to zoom or pan the model.

3. Select arc 2 and place the radius dimension on the left of the sketch. Modify the dimension value in the **Edit Dimension** edit box to **10** and press ENTER. The size of arc 10 is also modified because of the **Equal** constraint applied between these two entities.

4. Select line 5 and then line 7, and then place the dimension above the sketch. Modify the value of this dimension in the **Edit Dimension** edit box to **15** and press ENTER.

5. Select line 7 and place the dimension on the right of the sketch. Modify the value of this dimension in the **Edit Dimension** edit box to **20** and press ENTER.

6. Select the upper endpoint of line 7 and select line 1, and then place the dimension on the right of the previous dimension. Modify the value of this dimension to **40** and press ENTER. Next, exit the **Dimension** tool.

   With this step, all the dimensions have been applied to the sketch, refer to Figure 3-5, except the horizontal dimension between the center points of arcs 4 and 6 or arcs 8 and 6. The need of these dimensions depends on the constraints and dimensions assumed while drawing the sketch. If the sketch gets over-constrained, the **Autodesk Inventor Professional** message box is displayed. Choose **Cancel** from the message box. In this case, the dimension has already been assumed.

Drawing Circles

Once all the required dimensions and constraints have been applied to the sketch, you need to draw circles. Figure 3-6 indicates that circles are concentric with arcs 2 and 10.

1. Choose the **Center Point Circle** tool from the **Create** panel; you are prompted to select the center of the circle. Move the cursor close to the center of arc 2. Specify the center point when the cursor snaps to the center point of arc 2 and turns green. Next, move the cursor away from the center and specify a point to size the circle.

2. Similarly, draw the other circle taking the reference of the center of arc 10.

Adding Constraints to Circles

As both the circles have the same diameter, you can apply the Equal constraint to them. On applying the dimension to one of the circles, the other circle will automatically be forced to be created as per the specified diameter value as the Equal constraint has been applied on it.

1. Invoke the **Equal** constraint tool from the **Constrain** panel. Select the first circle and then the second circle to apply the Equal constraint.

Dimensioning Circles

1. Choose the **Dimension** tool from the **Constrain** panel and select the left circle. Place the dimension on the left of the sketch. In the **Edit Dimension** edit box, change the value of the diameter of the circle to **8** and press ENTER.

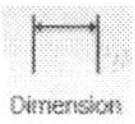

Notice that because of the Equal constraint, the size of the right circle is automatically modified to match the dimension of the left circle. The final sketch for Tutorial 1 after drawing and dimensioning circles is shown in Figure 3-6.

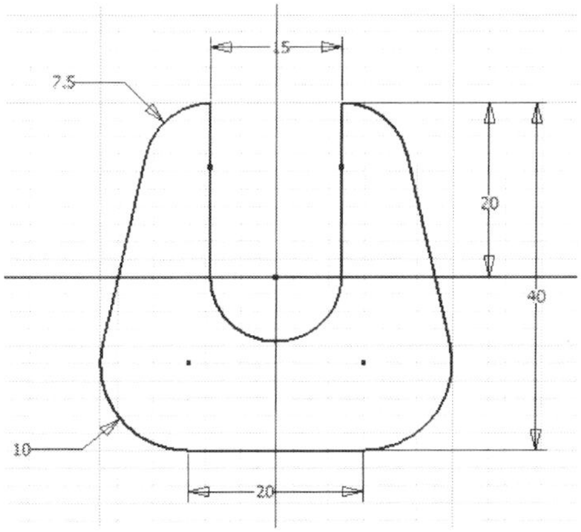

Figure 3-5 *Dimensioned sketch for Tutorial 3*

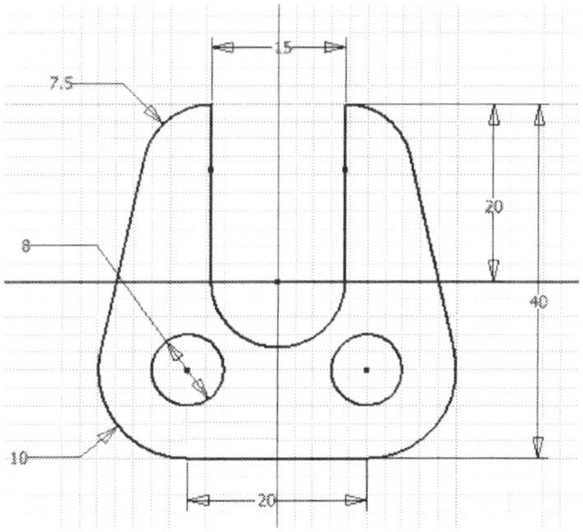

Figure 3-6 *The final dimensioned sketch for Tutorial 3*

Saving the Sketch

1. Choose the **Return** tool from the **Quick Access Toolbar** to exit the Sketching environment. Save this sketch with the name *Tutorial1* at the location given below: *C:\Inventor_2020/c03*

> **Note**
> *If the **Return** tool is not available in **Quick Access Toolbar**, you need to add this tool to the toolbar. To do so, choose the down arrow on the right of **Quick Access Toolbar**; a flyout is displayed. Next, choose the **Return** option from the flyout.*

2. Choose **Close > Close** from the **File** menu to close the file.

Tutorial 2

In this tutorial, you will draw the sketch of the model shown in Figure 3-7. The dimensions of the sketch are shown in Figure 3-8. After drawing the sketch, add constraints and then dimension it. The solid model is given for reference only. **(Expected time: 30 min)**

The following steps are required to complete this tutorial:

a. Start a new metric standard part file and invoke the Sketching environment.
b. Draw the outer loop of the sketch.
c. Add the required dimensions and constraints to the sketch.
d. Add the inner circle to the sketch and dimension it.
e. Save the sketch with the name *Tutorial2.ipt* and close the file.

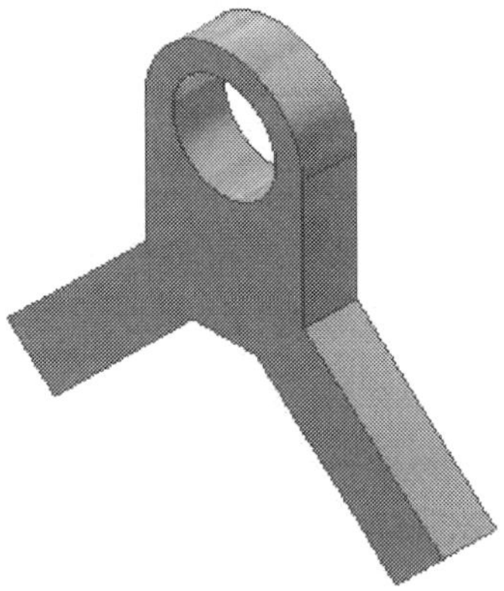

Figure 3-7 *Model for the sketch of Tutorial 2*

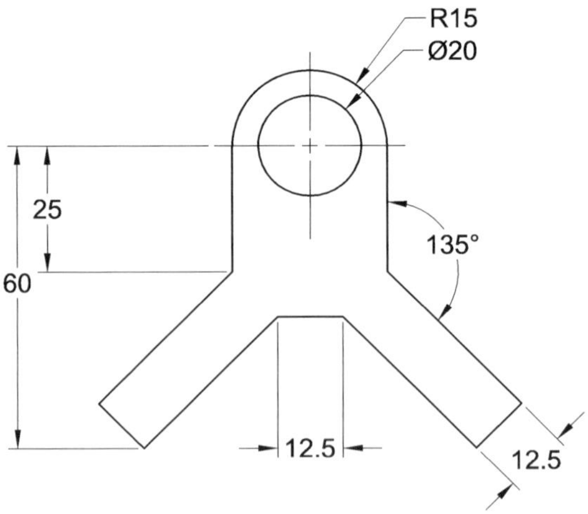

Figure 3-8 *The final dimensioned sketch for Tutorial 2*

Starting a New File and Invoking the Sketching Environment

1. Choose the **New** tool from the **Quick Access Toolbar** to display the **Create New File** dialog box. Start a new metric standard part file from the **Metric** tab of this dialog box.

2. Choose the **Start 2D Sketch** tool from the **Sketch** panel of the **3D Model** tab; the default planes are displayed and you are prompted to select the sketching plane.

3. Select the **XY Plane** as the sketching plane from the graphics window; the Sketching environment is invoked and the **XY Plane** becomes parallel to the screen.

Drawing the Outer Loop

1. Choose the **Line** tool from the **Create** panel of the **Sketch** tab to draw the outer loop, as shown in Figure 3-9. As mentioned earlier, you should draw the inner loop after drawing and dimensioning the outer loop. This is because once the outer loop is dimensioned, you can draw the inner loop by taking the reference of the outer loop.

   You can draw the arc while the **Line** tool is active. You can also use the temporary tracking option to draw this sketch. For your reference, the geometries in the sketch are numbered, see Figure 3-9.

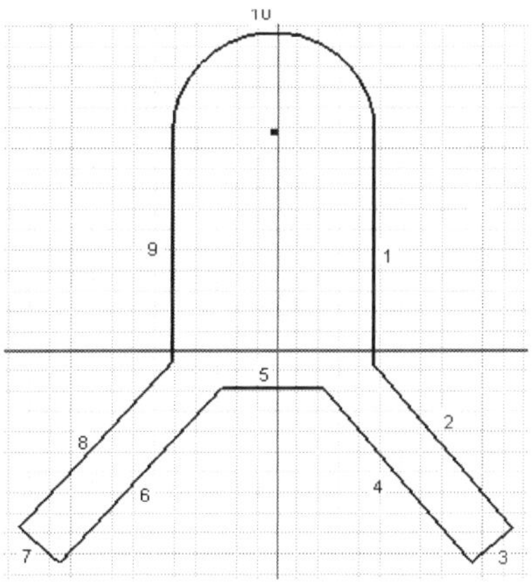

***Figure 3-9** Initial sketch with the geometries numbered*

Adding Constraints to the Outer Loop

1. Add the Equal constraint to lines 1 and 9, 2 and 8, 3 and 5, 5 and 7, and 4 and 6.

2. Add the Perpendicular Constraint to lines 2 and 3 and 7 and 8.

3. Add the Horizontal Constraint to the lower endpoints of lines 4 and 6.

4. Add the Tangent Constraint to lines 1 and 9 with arc 10, if it is missing.

Dimensioning the Outer Loop

1. Choose the **Dimension** tool from the **Constrain** panel of the **Sketch** tab; you are prompted to select the geometry to dimension. Select line 9 and place the dimension on the left of the sketch. Modify the dimension value in the **Edit Dimension** edit box to **25** and press ENTER.

2. Select the center of the arc and then the lower endpoint of line 6. Place the dimension on the left of the previous dimension. Modify the dimension value in the **Edit Dimension** edit box to **60** and press ENTER.

3. Select line 3 and then right-click to display the Marking menu. Choose **Aligned** from the Marking menu and then place the dimension below the sketch. Modify the dimension value in the **Edit Dimension** edit box to **12.5** and press ENTER.

   Notice that the length of lines 5 and 7 is also modified because of the Equal constraint.

4. Select lines 1 and 2 and then place the angular dimension on the right of the sketch. Modify the value of the angular dimension in the **Edit Dimension** edit box to **135** and press ENTER.

5. Select arc 10 and then place the radius dimension above the sketch. Modify the value of the radius of the arc in the **Edit Dimension** edit box to **15** and press ENTER.

   With this, the required dimensions have been applied to the outer loop. Even after adding all dimensions, the color of entities in the sketch remains green.

   Now, you can use the origin to fully constrain the initial sketch.

6. Choose the **Coincident Constraint** tool from the **Constrain** panel; you are prompted to select the first curve or point.

7. Select the center of the arc; you are prompted to select the second curve or point.

8. Select the origin. The entire sketch shifts itself such that the center of arc of the sketch is at the origin. After the sketch is shifted to a new place, it may not be visible completely in the drawing window.

9. Choose the **Zoom All** tool from the **Navigation Bar > Zoom** flyout to fit the sketch into the drawing window. You will notice that all the entities in the sketch turn purple, indicating that the sketch is fully constrained. Press the ESC key to exit the **Coincident Constraint** tool.

Drawing the Circle

1. Choose **Center Point Circle** from the **Create** panel of the **Sketch** tab; you are prompted to select the center of the circle.

2. Move the cursor close to the center of the arc; the cursor snaps to the center point and turns green. Select this point as the center of the circle and then move the cursor away from the center to size the circle. Specify a point to give it an approximate size.

Dimensioning the Circle

1. Choose **Dimension** from the **Constrain** panel of the **Sketch** tab and select the circle. Place the diameter dimension below the arc dimension. Enter **20** in the **Edit Dimension** edit box and then press ENTER. This completes the sketch for Tutorial 4. The final dimensioned sketch is shown in Figure 3-10.

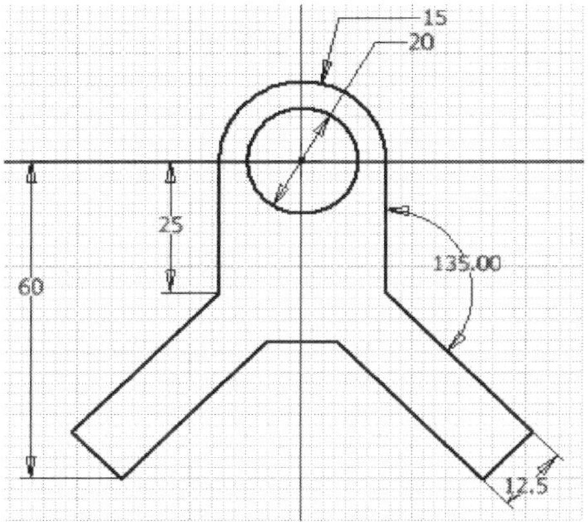

Figure 3-10 *The final dimensioned sketch for Tutorial 2*

Saving the Sketch

1. Choose the **Finish Sketch** option from the **Exit** panel of the **Sketch** tab to exit the sketching environment. Alternatively, choose the **Finish 2D Sketch** option from the Marking menu to exit the sketching environment

2. Save the sketch with the name *Tutorial2* at the location given below:

   C:\Inventor_2020\c03

3. Choose **Close > Close** from **File** menu to close the file.

EXERCISES

Exercise 1

Draw the sketch of the model shown in Figure 3-11. The sketch to be drawn is shown in Figure 3-12. After drawing the sketch, add the required constraints to it and then dimension it. **(Expected time: 30 min)**

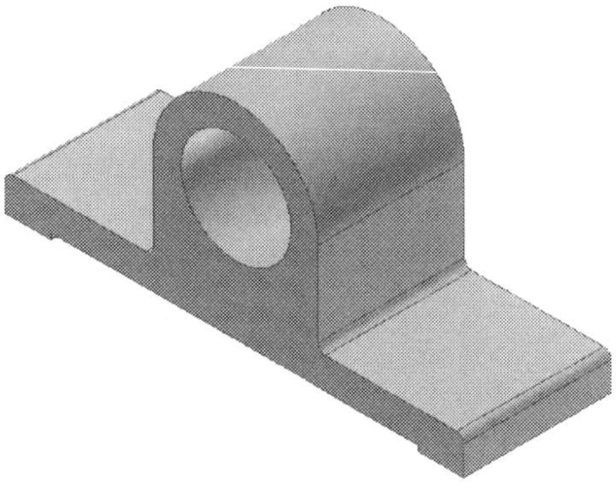

Figure 3-11 Model for Exercise 1

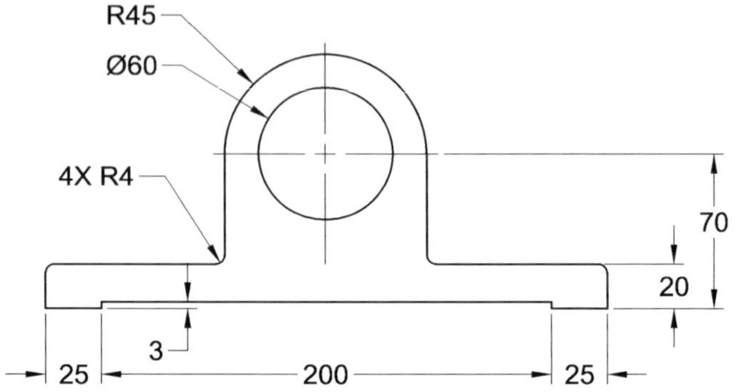

Figure 3-12 Sketch for Exercise 1

Exercise 2

Draw the sketch of the model shown in Figure 3-13. The sketch to be drawn is shown in Figure 3-14. After drawing the sketch, add the required constraints to it and then dimension it.

(Expected time: 30 min)

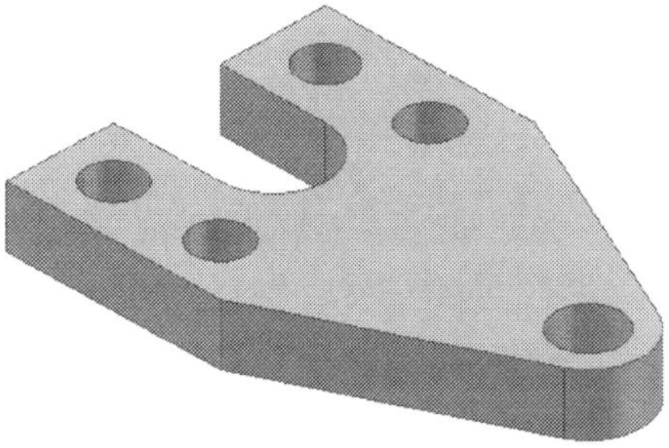

Figure 3-13 Model for Exercise 2

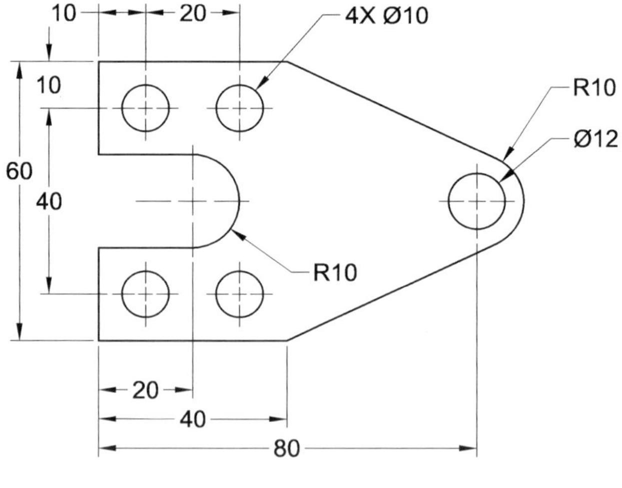

Figure 3-14 Sketch for Exercise 2

Chapter 4

Editing, Extruding, and Revolving the Sketches

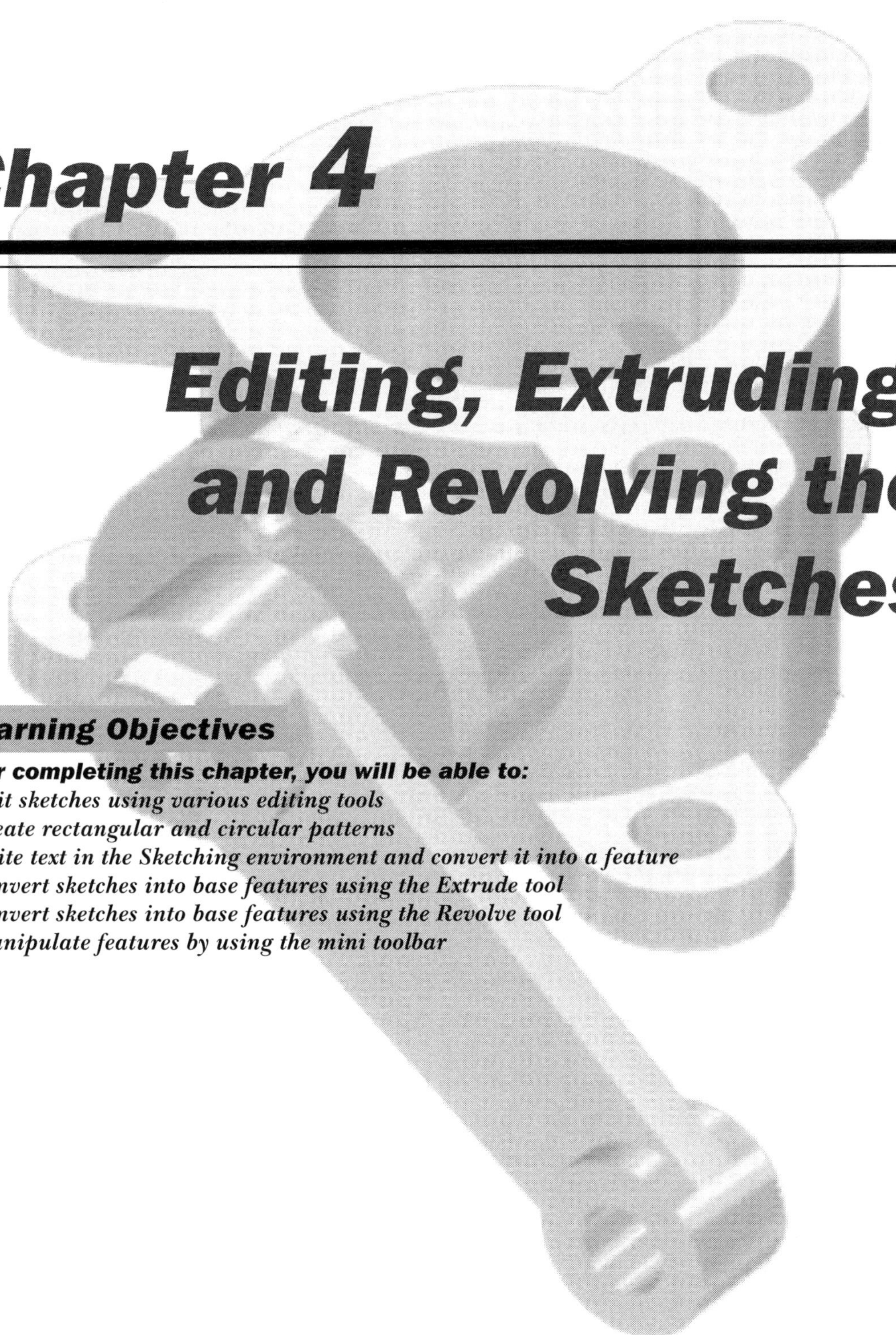

Learning Objectives

After completing this chapter, you will be able to:

- *Edit sketches using various editing tools*
- *Create rectangular and circular patterns*
- *Write text in the Sketching environment and convert it into a feature*
- *Convert sketches into base features using the Extrude tool*
- *Convert sketches into base features using the Revolve tool*
- *Manipulate features by using the mini toolbar*

EDITING SKETCHED ENTITIES

Autodesk Inventor provides you with a number of tools that can be used to edit the sketched entities. These tools are discussed next.

Extending Sketched Entities

Ribbon: Sketch > Modify > Extend

You can extend or lengthen the selected entity up to a specified boundary by using the **Extend** tool. For using this tool, you should have at least two entities such that when extended, they meet at a point. Taking the reference of one of the entities, the other will be extended. The entities that can be extended using this tool are lines, splines, and arcs.

Trimming Sketched Entities

Ribbon: Sketch > Modify > Trim

In Inventor, you can cut the length of an entity. You can do so by using the **Trim** tool. This tool chops the selected sketched entity by using an edge (also called the knife-edge). The knife-edge, in its current form, may or may not actually intersect the entity to be trimmed. However, when extended, the knife-edge must intersect the entity to be trimmed.

Splitting Sketched Entities

Ribbon: Sketch > Modify > Split

The **Split** tool is used to break a sketched entity into two or more entities at the intersection point(s) with another sketched entity.

Offsetting Sketched Entities

Ribbon: Sketch > Modify > Offset

Offsetting is one of the easiest methods of drawing parallel lines, concentric arcs and circles. You can select the entire loop as a single entity or select the individual entities to be offset.

Mirroring Sketched Entities

Ribbon: Sketch > Pattern > Mirror

The **Mirror** tool is used to create mirror images of the selected entities. The entities are mirrored about a straight line segment. This tool is used to draw sketches that are symmetrical about a line or sketches.

Moving Sketched Entities

Ribbon: Sketch > Modify > Move

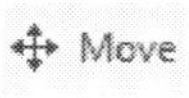

The **Move** tool is used to move one or more selected sketched entities from one point to the other. The points that can be used to move the entities are the sketched points, the endpoints of lines, arcs, splines, and the center points of arcs, circles, and ellipses.

Rotating Sketched Entities

Ribbon: Sketch > Modify > Rotate

The **Rotate** tool is used to rotate the selected sketched entities about a specified center point. You can also use this tool to create a copy of the selected entities while rotating them.

CREATING PATTERNS

Generally, in the mechanical industry, you come across various designs that consist of multiple copies of a sketched feature arranged in a particular fashion. For example, it can be multiple grooves around an imaginary circle. It can also be along the edges of an imaginary rectangle, such as the grooves in the pedestal bearing. Drawing the sketches for such features again and again is a very tedious and time-consuming process. To avoid this lengthy process, Autodesk Inventor provides you with an option for creating patterns of the sketched entities during the sketching stage itself. The patterns are defined as the sequential arrangement of the copies of the selected entities. You can create the patterns in a rectangular or circular fashion. Both these types of patterns are discussed next.

Creating Rectangular Patterns

Ribbon: Sketch > Pattern > Rectangular Pattern

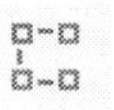

Rectangular patterns are the patterns that arrange the copies of the selected entities in rows and columns.

Creating Circular Patterns

Ribbon: Sketch > Pattern > Circular Pattern

Circular patterns are the patterns created around the circumference of an imaginary circle. To create the circular pattern, you will have to define the center of that imaginary circle.

INSERTING IMAGES AND DOCUMENTS IN SKETCHES

Ribbon: Sketch > Insert > Insert Image

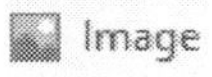

The **Insert Image** tool allows you to insert the external images such as JPG, BMP, PNG, GIF, and so on in the sketch. You can also insert Word documents or Excel spreadsheets using this tool.

TOLERANCES

In simple words, tolerance is defined as a permissible variation from the actual value. As it is an allowed variation, you can vary the dimension of the component through the specified value while manufacturing.

CONVERTING THE BASE SKETCH INTO A BASE FEATURE

As mentioned earlier, any 3D design is a combination of various sketched, placed, and work features. The first feature, generally, is a sketched feature. You have already learned to draw the sketches and to dimension them. After you have finished drawing and dimensioning the sketch, choose the **Finish Sketch** button from the **Exit** panel of the **Sketch** tab. On choosing this button, you will exit the sketching environment and enter the **Part** module. You will also notice that the **Sketch** tab is replaced by the **3D Model** tab. Autodesk Inventor provides you with a number of tools such as **Extrude**, **Revolve**, **Loft**, **Sweep**, and so on to convert these sketches into base features. However, in this chapter, you will learn the use of the **Extrude** and **Revolve** tools for converting the sketch into a base feature. The remaining tools will be discussed in the later chapters.

EXTRUDING THE SKETCH

Ribbon:	3D Model > Create > Extrude

The **Extrude** tool is one of the most extensively used tools for creating a design. Extrusion is a process of adding or removing material defined by a sketch, normal to the current sketching plane.

REVOLVING THE SKETCH

Ribbon:	3D Model > Create > Revolve

The **Revolve** tool is used to create circular features like shafts, couplings, pulleys, and so on. You can also use this tool for creating cylindrical cut features. A revolved feature is created by revolving the sketch about an axis.

Adding Tolerances to the Dimensions in the Sketching Environment

In Autodesk Inventor, tolerances are added to the sketch after dimensioning it. To add tolerance to a dimension, right-click on the dimension text and choose **Dimension Properties** from the Marking menu; the **Dimension Properties** dialog box will be displayed. You can use the options in the **Dimension Settings** tab of this dialog box to add tolerances, refer to Figure 4-1.

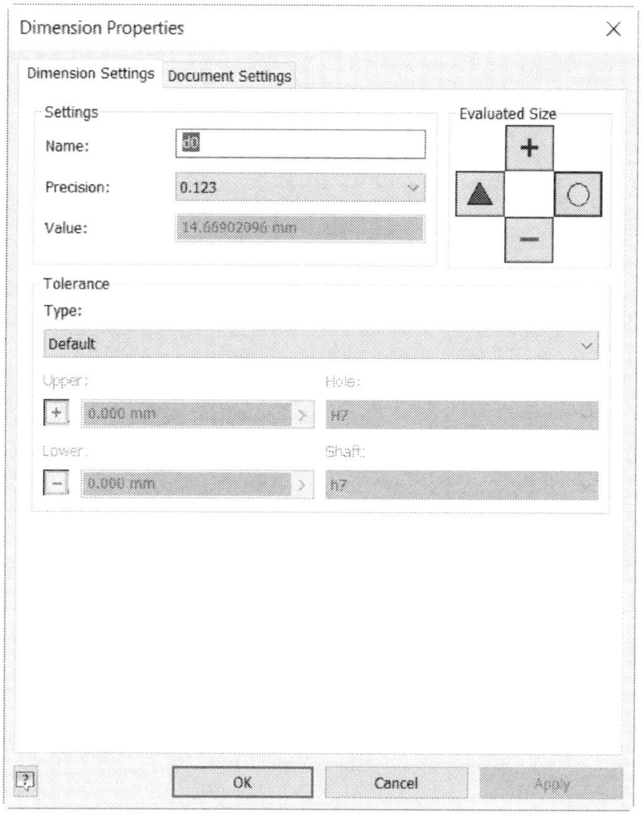

Figure 4-1 *The **Dimension Settings** tab of the **Dimension Properties** dialog box*

MANIPULATE FEATURES BY USING THE MINI TOOLBAR

The manipulators appear in the form of arrows. The manipulators for the **Extrude**, **Revolve**, and **Hole** tools are represented by a straight arrow, a curved arrow, and a sphere, respectively. With the help of these manipulators, you can specify the extrusion depth of an extruded feature, angle of revolution for a revolved feature, and the location of the hole dynamically.

If a command is not active and you select an edge, the mini toolbar will provide you with options to create a fillet or a chamfer feature, refer to Figure 4-2. Similarly, if you select a sketch without invoking any command, the mini toolbar displays the options to create extruded feature, revolved feature, and hole as well as the option to edit a sketch, refer to Figure 4-3.

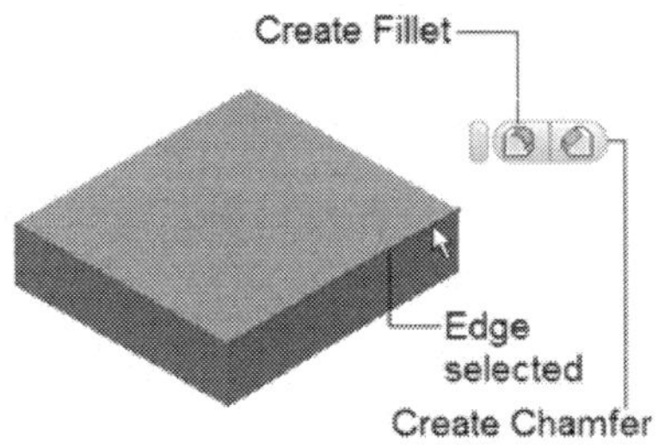

Figure 4-2 Mini toolbar displayed on selecting an edge

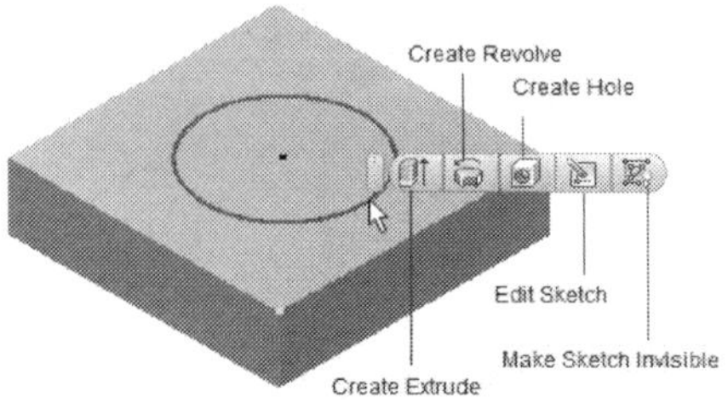

Figure 4-3 Mini toolbar displayed on selecting a sketch

CREATING FREEFORM SHAPES

The Solid primitive freeform shapes form the basic building blocks of a complex solid. In Autodesk Inventor, you can directly create the primitive freeform shapes such as Box, Plane, Cylinder, Sphere, Torus, and Quadball.

CREATING PREDEFINED SOLID PRIMITIVES

The Solid primitives form the basic building blocks for a complex solid. Autodesk Inventor has four predefined solid primitives that can be used to construct a solid model such as Box, Cylinder, Sphere, and Torus.

TUTORIALS

Tutorial 1

In this tutorial, you will create the model shown in Figure 4-4. Its dimensions are shown in Figure 4-5. The extrusion height for the model is 10 mm. After extruding it, you will set the option to cast the X-ray ground shadow. **(Expected time: 45 min)**

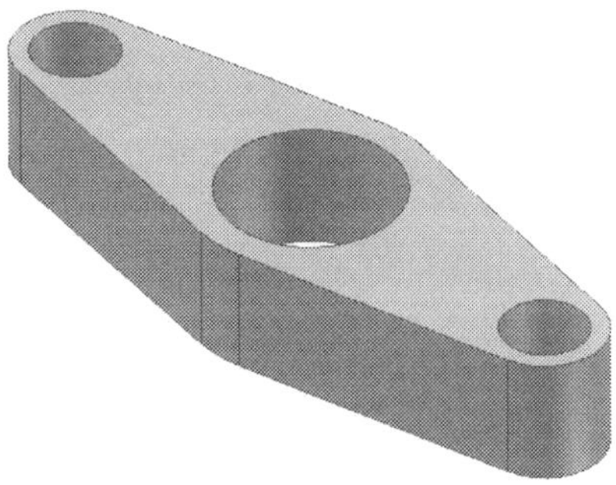

Figure 4-4 Model for Tutorial 1

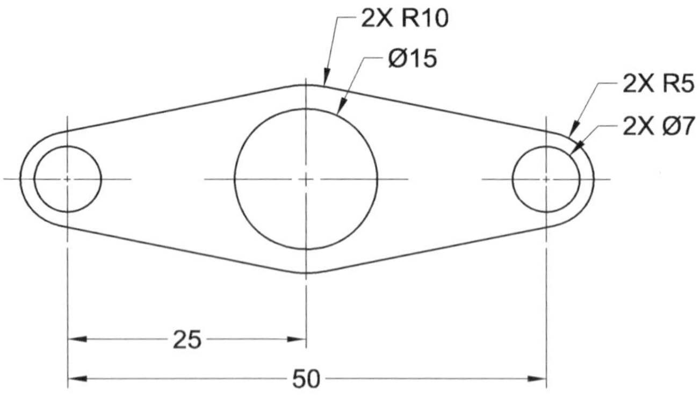

Figure 4-5 Dimensions of the model

The following steps are required to complete this tutorial:

a. Start a new metric standard part file. Draw the sketch of the outer loop and add constraints to it.
b. Draw the inner circles and add the required constraints. Dimension the complete sketch.
c. Extrude the sketch to a distance of 10 mm using the **Extrude** tool.
d. Cast the object and ambient shadows on the final model.

Starting a New Part File

If you have installed Autodesk Inventor with millimeter as the unit of measurement, you can directly start a new metric standard part file, thus avoiding the use of the **Create New File** dialog box for opening a new part file.

1. Choose the arrow on the right of the **New** tool from the **Quick Access** toolbar; a flyout is displayed.

2. Choose the **Part** option from this flyout to start a new metric part file. If Autodesk Inventor was not installed with millimeter as the measurement unit, you need to choose the **Metric** tab of the **Create New File** dialog box to start the new metric standard part file.

Note

*You can change the units used in Inventor file by using the **Document Settings** dialog box. To invoke this dialog box, choose the **Document Settings** tool from the **Options** panel in the **Tools** tab. Next, choose the **Units** tab in this dialog box to display various options related to units. Select the required unit from the **Length** drop-down list in the **Units** area of this dialog box. Next, choose **Apply** and then **Close** to exit the dialog box.*

Creating the Sketch of the Model

As shown in Figure 4-5, the sketch is a combination of an outer loop and three circles. First, you will create the outer loop. This outer loop will be created by drawing three circles, two at the ends and one at the center, and then connecting the middle circle with the other two circles through tangent lines. Finally, you will trim the unwanted portions of the circles.

1. Choose the **Start 2D Sketch** button from the **Sketch** panel of the **3D Model** tab; the default planes are displayed and you are prompted to select a sketching plane.

2. Now, select the **XZ Plane** from the **Browser Bar** as the sketching plane; the Sketching environment is invoked and the **XZ Plane** becomes parallel to the screen. Alternatively, you can also select **XZ Plane** from the graphics window.

3. At this position rotate the ViewCube at 90 degrees in the anticlockwise direction. Next, click on the down arrow available next to the ViewCube; a flyout is displayed. Next, choose **Set Current View as >Top** from the flyout.

4. Draw the sketch, which is a combination of three circles and tangent lines. Add the **Tangent** constraint to the lines wherever it is missing. Also, add the **Equal** constraint to all four lines, and the circles on the left and right sides. Finally, add the **Horizontal** constraint to the centers of the circles. The sketch, after drawing and adding constraints, should look similar to the one shown in Figure 4-6.

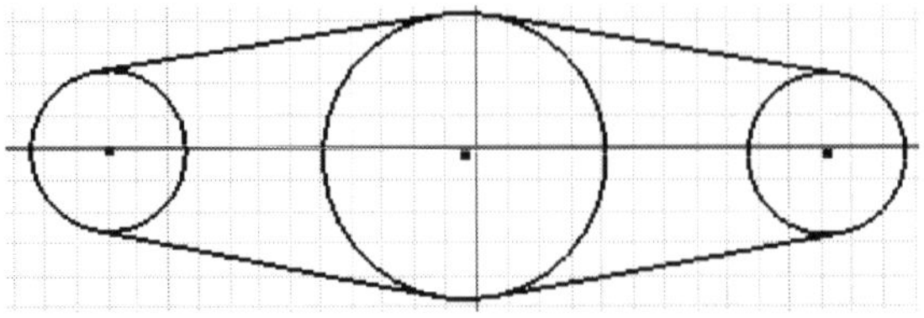

Figure 4-6 *Sketch after drawing and adding constraints*

Next, you need to remove the unwanted portions of the circles using the **Trim** tool.

5. Choose the **Trim** tool from the **Modify** panel of the **Sketch** tab; you are prompted to select the portion of the curves to be trimmed.

6. Move the cursor close to the right half of the left circle.

   As you move the cursor close to the circle, the color of the circle turns white and it appears as dashed on the right side.

7. Specify a point on the right half of the left circle; the right half of this circle is trimmed. Similarly, select portions of the other circles to trim, as shown in Figure 4-7.

8. Next, draw three circles concentric to the three trimmed arcs. Add the **Equal** constraint between the left and right circles. The sketch after drawing the circles and applying the **Equal** constraint to them should look similar to the one shown in Figure 4-8.

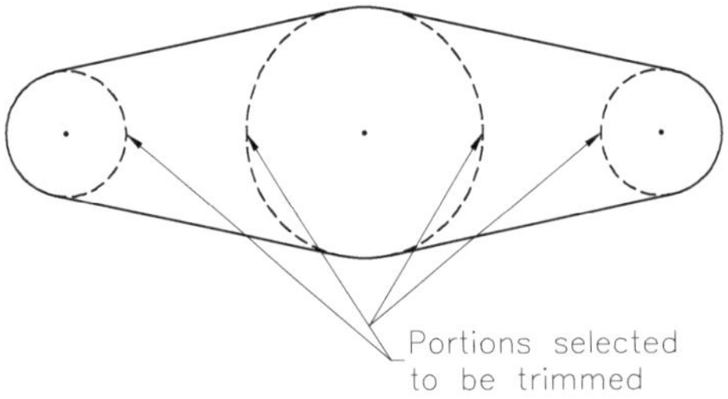

Figure 4-7 *Selecting the portions to be trimmed*

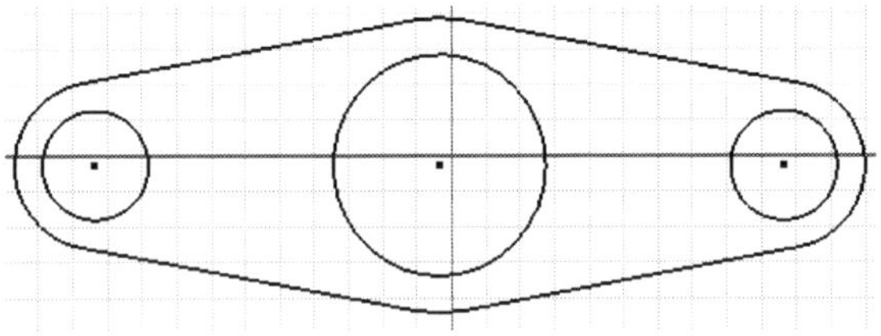

Figure 4-8 *Sketch after creating the outer loop and the three circles*

Dimensioning the Sketch

1. Dimension the sketch as required, refer to Figure 4-5. The sketch, after it has been dimensioned, will look similar to the one shown in Figure 4-9.

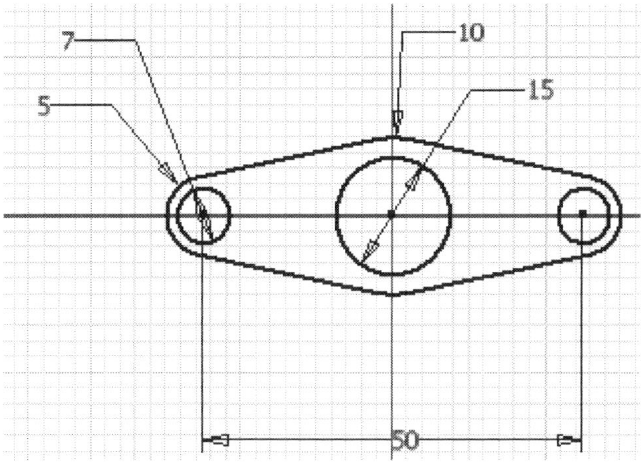

Figure 4-9 Sketch after adding dimensions

Extruding the Sketch

The sketch consists of four loops: the outer loop and the three circles. When you extrude this sketch, the three circles will be automatically subtracted from the outer loop. As a result, you will get the required model. However, this is possible only if the point specified for selecting the profile is inside the outer loop but outside all three circles.

1. Choose the **Finish Sketch** button from the **Exit** panel of the **Sketch** tab to exit the sketching environment. Alternatively, choose the **Finish 2D Sketch** option from the Marking Menu that is displayed on right-clicking anywhere in the graphics window. You will notice that the current view is changed to the Home view.

2. Choose the **Extrude** tool from the **Create** panel of the **3D Model** tab; the **Properties-Extrude** dialog box is invoked. As the sketch consists of more than one loop, you need to select the profile to be extruded.

3. Move the cursor to a point anywhere inside the outer loop but outside all three circles; the profile is selected and highlighted. Notice that the area inside any of these circles is not shaded. This shows that the area inside these circles will not be extruded. This is also one of the methods to cross check whether the profile selected is the one you need to extrude or not.

4. Click inside the shaded profile; the preview of the model is displayed in the graphics window.

5. Accept the default values and then choose **OK** from **Properties-Extrude** dialog box to extrude the profile to a depth of 10 mm.

   You may need to change the camera type from perspective to orthographic if the camera type currently used is perspective. To change the camera type, choose the **Orthographic** button from the **Orthographic** drop-down in the **Appearance** panel of the **View** tab. The final model is shown in Figure 4-10.

Casting the Object Shadow and the Ambient Shadow

Next, you need to apply the object shadow and ambient shadow to the model. You can cast these shadows using the options in the **Shadows** drop-down list in the **Appearance** panel of the **View** tab.

1. Click on the down arrow on right of **Shadows** in the **Appearance** panel; a drop-down is displayed.

2. In this drop-down, the **Ambient Shadows** check box is selected by default. Select the **Object Shadows** check box from the drop-down; the object and ambient shadows are cast on the model, as shown in Figure 4-11.

Saving the Model

1. Choose the **Save** tool from the **Quick Access Toolbar**; the **Save As** dialog box is displayed.

2. Save the model with the name *Tutorial1* at the location given below:
   C:\Inventor_2020\c04

3. Choose **Close > Close** from the **File** menu to close the file.

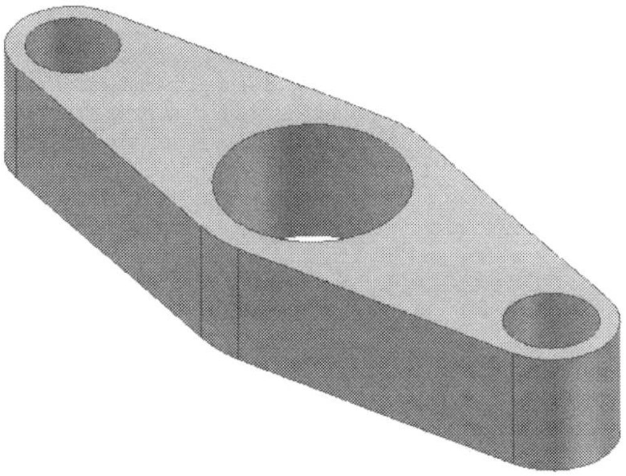

Figure 4-10 *Final model for Tutorial 1*

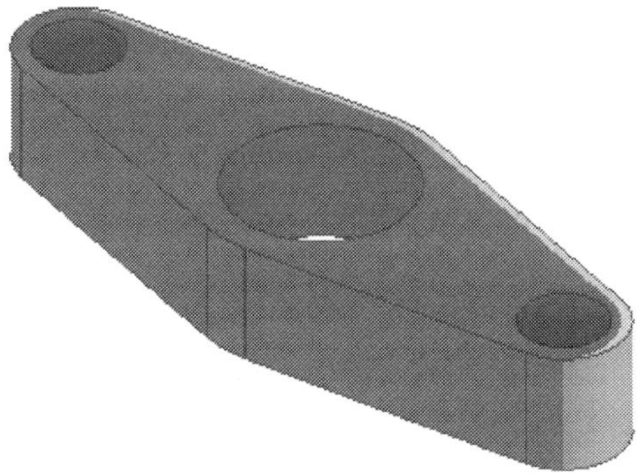

Figure 4-11 *Object and ambient shadows applied on the model*

Tutorial 2

In this tutorial, you will extrude the text and then change the visual style of the extruded text, as shown in Figure 4-12. The font size of the text is 5 mm and the height of extrusion of the text is 2.5 mm. **(Expected time: 45 min)**

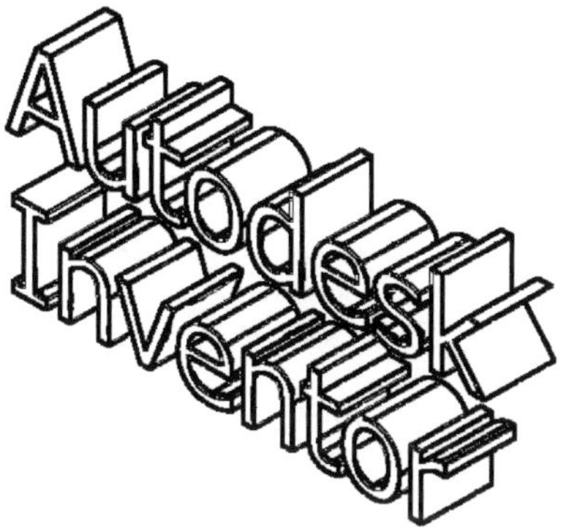

Figure 4-12 *Extruded text with the hand drawn appearance*

The following steps are required to complete this tutorial:

a. Start a new metric standard part file and write the text in the sketching environment.
b. Extrude the text to a distance of 2.5 mm using the **Extrude** tool.
c. Change the visual style of the extruded text.

Starting a New File and Writing the Text

1. Start a new metric standard part file using the **Metric** tab of the **Create New File** dialog box. Also, ensure that the display of the shadow is turned off.

2. Choose the **Start 2D Sketch** button from the **Sketch** panel of the **3D Model** tab; the default planes are displayed and you are prompted to select the sketching plane.

3. Now, select the **XY Plane** as the sketching plane from the **Browser Bar**; the Sketching environment is invoked and the **XY Plane** becomes parallel to the screen. Alternatively, you can also select the **XY Plane** from the Graphics window.

4. Choose the **Text** tool from the **Create** panel of the **Sketch** tab; you are prompted to click on the location of the text.

5. Specify a point anywhere in the drawing window; the **Format Text** dialog box is displayed.

6. Select the **Arial** font from the **Font** drop-down list and then set the font height to 5 mm in the **Size** edit box.

7. Type the text **Autodesk Inventor** in the Text Window in two lines. Choose **OK** to exit the dialog box; the typed text is displayed in the drawing window, as shown in Figure 4-13.

Extruding the Text

Next, you need to exit the sketching environment and extrude the text.

1. Choose the **Finish Sketch** button from the **Exit** panel of the **Sketch** tab and exit the sketching environment. Alternatively, choose the **Finish 2D Sketch** option from the Marking Menu that is displayed when you right-click anywhere in the graphics window. Note that the current view is changed to the home or isometric view.

2. Choose the **Extrude** tool from the **Create** panel of the **3D Model** tab; the **Properties-Extrude** dialog box is invoked, and you are prompted to select the profile to extrude.

3. Move the cursor over the text and select it when it is highlighted.

4. Specify **2.5** as the extrusion depth in the **Distance A** edit box. Choose **OK** from the **Properties-Extrude** dialog box; the text is extruded to a distance of 2.5 mm, refer to Figure 4-14.

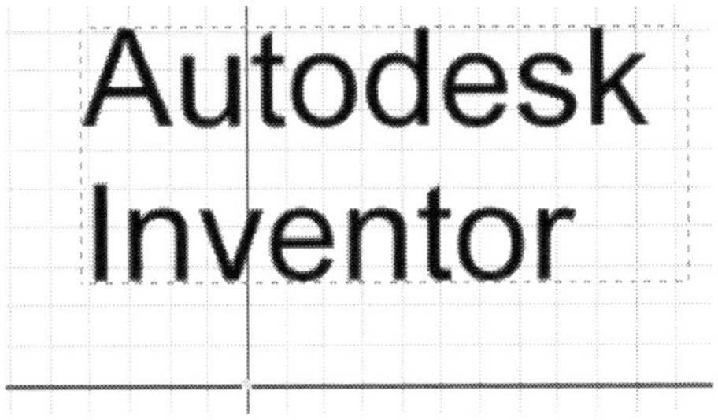

Figure 4-13 *Text in the Sketching environment*

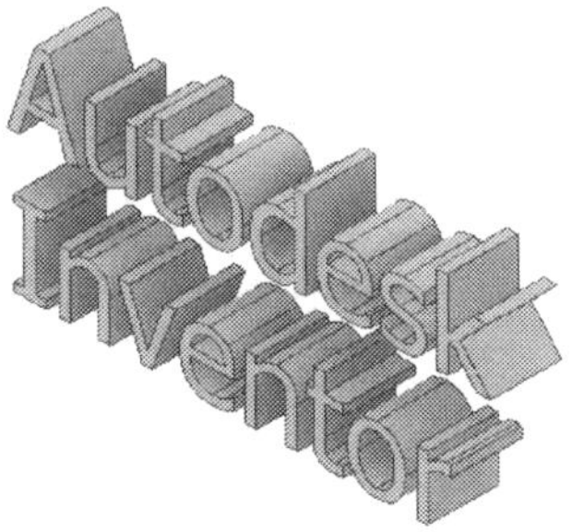

Figure 4-14 *Text after extrusion*

Changing the Visual Style

After extruding the text, you need to change its visual style.

1. Click the **Visual Style** in the **Appearance** panel of the **View** tab; a drop-down is displayed.

2. Choose the **Sketch Illustration** tool from this drop-down to give the extruded text a hand-drawn appearance. The final extruded text with the hand-drawn appearance is shown in Figure 4-15.

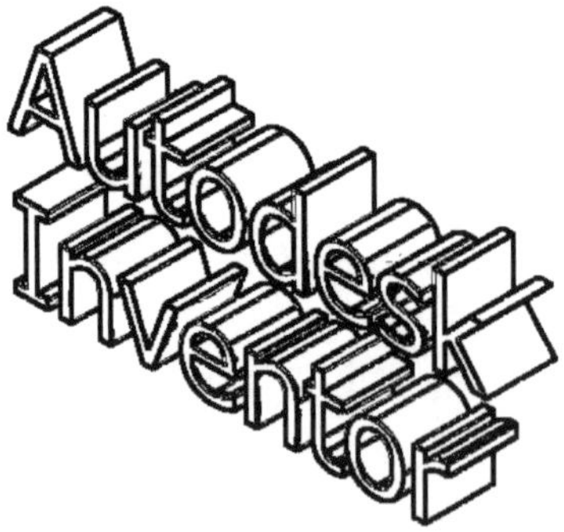

Figure 4-15 *Text after changing the visual style*

Saving the Sketch

1. Choose the **Save** tool from the **Quick Access Toolbar**; the **Save As** dialog box is displayed. Save the model with the name *Tutorial2* at the location given below:

C:\Inventor_2020\c04

2. Choose **Close > Close** from the **File** menu to close the file.

EXERCISES

Exercise 1

In this exercise, you will extrude the sketch drawn in Exercise 2 of Chapter 3, refer to Figure 4-16. The extrusion depth for the model is 15 mm. **(Expected time: 30 min)**

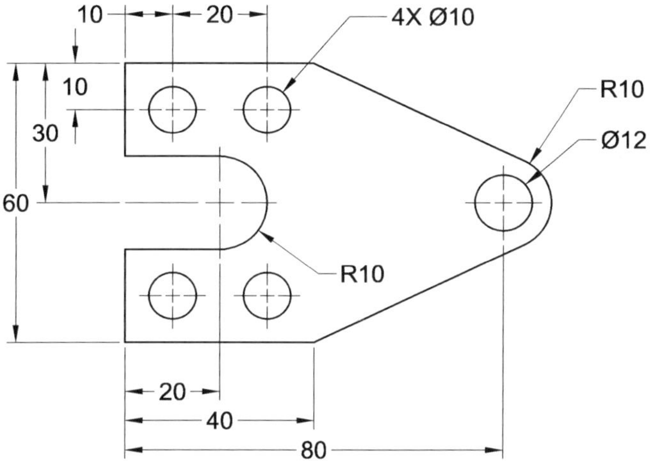

Figure 4-16 *Sketch for Exercise 1*

Exercise 2

In this exercise, you will extrude the sketch drawn in Exercise 1 of Chapter 3, refer to Figure 4-17. The extrusion depth for the model is 80 mm. **(Expected time: 30 min)**

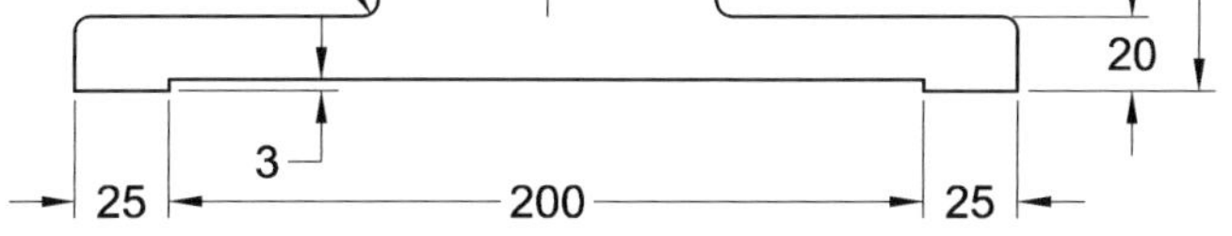

Figure 4-17 *Sketch for Exercise 2*

This page is intentionally left blank

Chapter 5

Other Sketching and Modeling Options

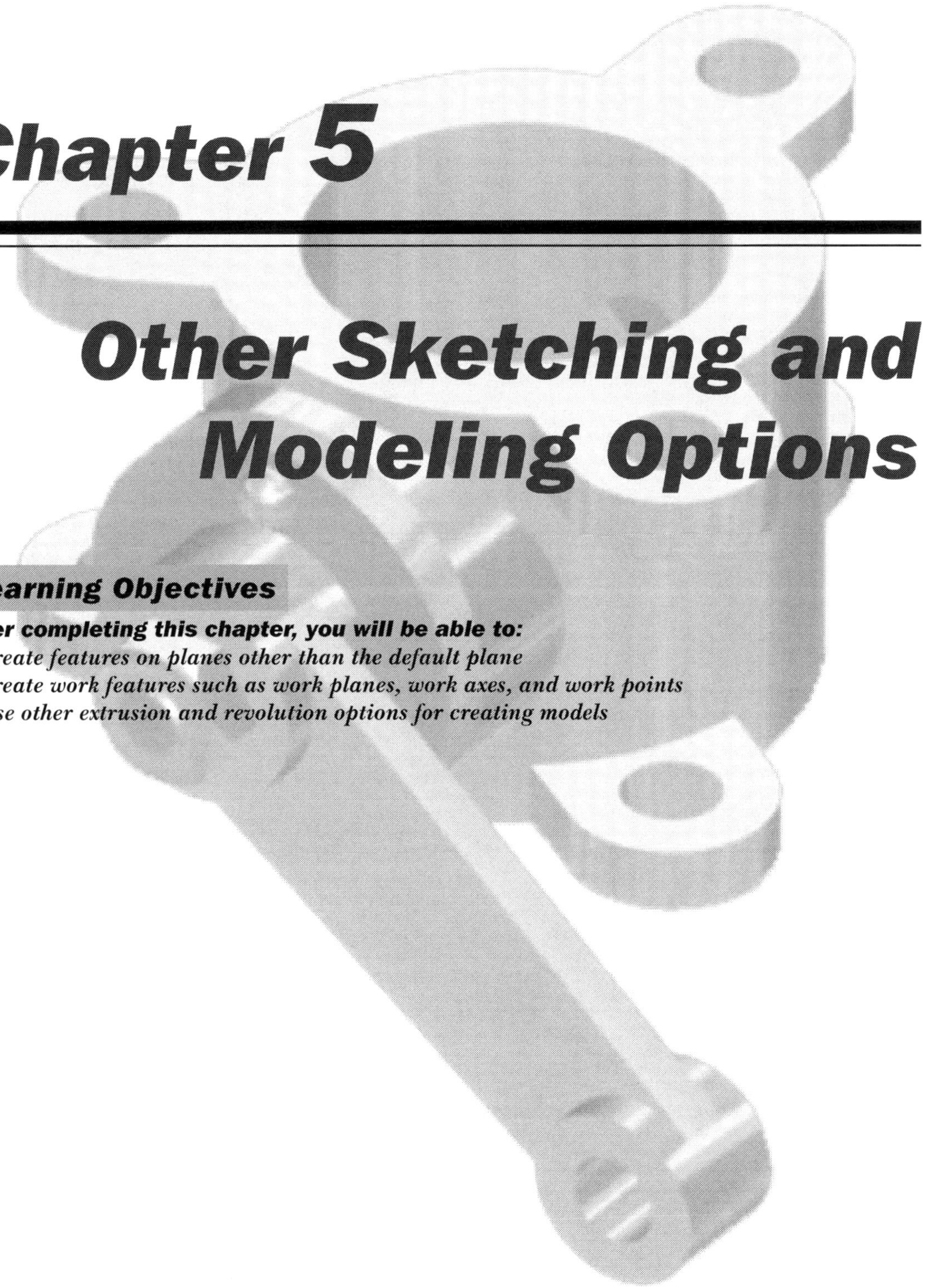

Learning Objectives

After completing this chapter, you will be able to:

- *Create features on planes other than the default plane*
- *Create work features such as work planes, work axes, and work points*
- *Use other extrusion and revolution options for creating models*

CREATING FEATURES ON PLANES OTHER THAN THE DEFAULT PLANES

In the earlier chapters, you created basic models by extruding or revolving the sketches. All those models were created on a single sketching plane, either XY, YZ, or XZ plane. But most mechanical designs consist of multiple sketched features, referenced geometries, and placed features. These features are integrated together to complete a model. Most of these features lie on different planes. When you start a new Autodesk Inventor part file and try to invoke a sketching environment, you are prompted to select the plane on which you want to draw the sketch. On the basis of design requirements, you can select any plane to create the base feature. To create additional sketched features, you need to select an existing plane or a planar surface, or you need to create a plane that will be used as a sketching plane. For example, consider the model shown in Figure 5-1.

The base feature for this model is shown in Figure 5-2. The sketch for the base feature is drawn on the XZ plane. As mentioned earlier, after creating the base feature, you need to create other sketched features, placed features, and work features, refer to Figure 5-3. The extrude features shown in Figure 5-3 require additional sketching planes on which the sketch for the other features will be created.

It is evident from Figure 5-3 that additional features created on the base feature do not lie on the same sketching plane. They are created by defining additional sketching planes. Also, appropriate extrusion options are selected at the time of creating these features.

Figure 5-1 *Model created by combining various features*

Figure 5-2 *Base feature for the model*

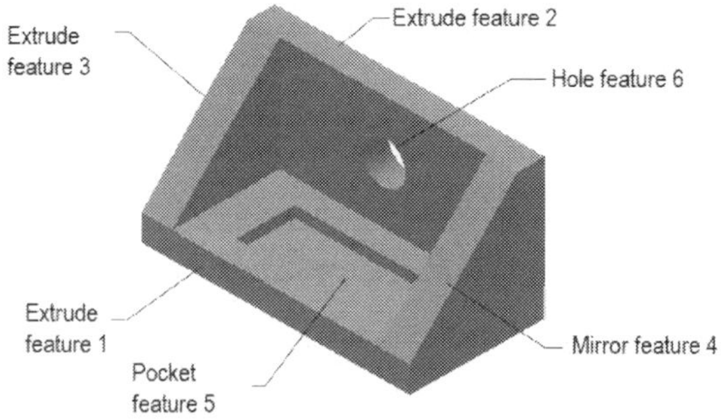

Figure 5-3 *Model after adding other features*

WORK FEATURES

Work features are parametric features that are associated with a model. Autodesk Inventor has provided three types of work features to assist you in creating a design. The three types of work features are:

- Work Planes
- Work Axes
- Work Points

OTHER EXTRUSION OPTIONS

The **Properties-Extrude** dialog box cannot be invoked until the sketch of base feature is created in the Sketching environment. Once you create the base feature in the Sketching environment and choose the **Finish 2D Sketch** option from the Marking menu, you will automatically be taken to the **3D Model** tab of the **Ribbon**. If you choose the **Extrude** tool from the **Create** panel of this tab, the **Properties-Extrude** dialog box will be displayed.

Some of the options of the **Properties-Extrude** dialog box will not be available until you have created the base feature. Once the base feature is created, all the options in this dialog box will be available. These options are discussed next.

Join

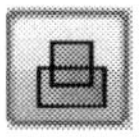

This button is used to create an extruded feature by adding a new material to an existing feature. This button will be available only after you have created the base feature.

Cut

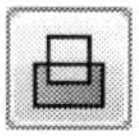

This button will be available only after you have created the base feature and is used to create an extruded feature by removing material from an existing feature.

Intersect

This button is used to create an extruded feature by using the material that is common to both the existing feature and the sketch.

New Solid

On choosing this button, a new body will be created. The new body will be independent of the existing body and will be listed in the **Solid Bodies** node of the **Browser Bar**.

OTHER REVOLVE OPTIONS

When you create a base feature, the options such as **Join**, **Cut**, **Intersect**, and **New Solid** will be available in the **Properties-Revolve** dialog box. The functions of these options are same as that explained in the **Extrude** tool section.

TUTORIALS

Tutorial 1

In this tutorial, you will create the model of the Standard Bracket shown in Figure 5-4. The views and dimensions of the model are shown in the same figure.

(Expected time: 30 min)

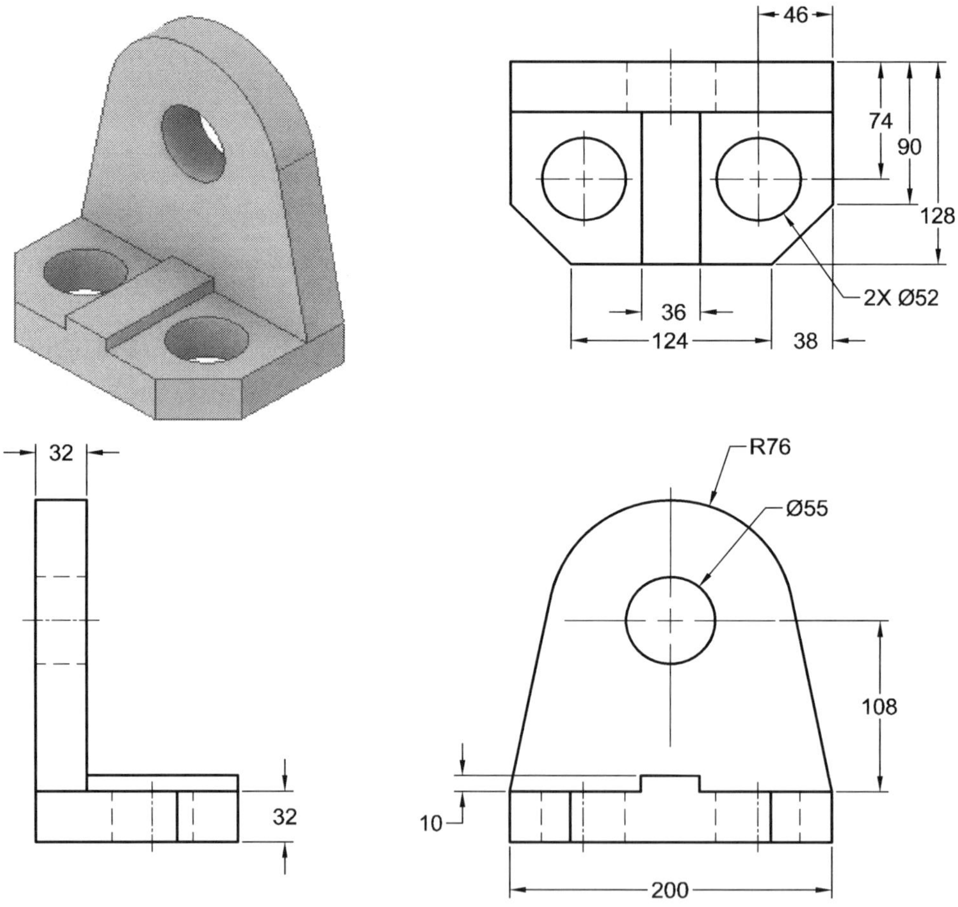

Figure 5-4 *Views and dimensions for Tutorial 1*

The following steps are required to complete this tutorial:

a. On the XZ plane, create the base feature with two holes.
b. Define a new sketch plane on the back face of the base feature and create the join feature with a hole.
c. Define a new sketch plane on the front face of the model and create the rectangular join feature.

Creating and Dimensioning the Sketch of the Base Feature

1. Start Autodesk Inventor and then start a new metric standard part file.

2. Choose the **Start 2D Sketch** button from the **Sketch** panel of the **3D Model** tab; the default planes are displayed and you are prompted to select the sketch plane.

3. Select the **XZ** plane as the sketching plane from the **Browser Bar**; the Sketching environment is invoked and the **XZ** plane becomes parallel to the screen.

4. At this position, rotate the ViewCube at 90 degrees in the anticlockwise direction. Next, click on the down arrow available next to the ViewCube; a flyout is displayed. Next, choose **Set Current View as >Top** from the flyout.

5. Draw the sketch of the base feature using various sketching tools, see Figure 5-5.

6. Add required constraints and dimensions to the sketch to make it fully constrained.

7. Choose the **Finish Sketch** button from the **Exit** panel of the **Sketch** tab; you will exit the Sketching environment and the current view is changed to the home view or isometric view.

Extruding the Base Sketch

After creating the sketch, you need to extrude it to create the base feature.

1. Using the **Extrude** tool, extrude the sketch upto a distance of 32 mm.

   As the sketch has multiple loops, you need to specify the profile to be extruded. Make sure you define the profile to be extruded by specifying a point outside the circles but inside the outer loop and choose **OK**. The model after creating the base feature is shown in Figure 5-6.

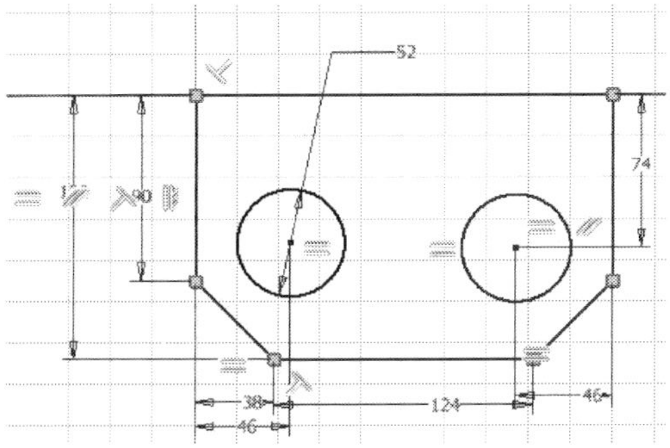

Figure 5-5 *Sketch of the base feature*

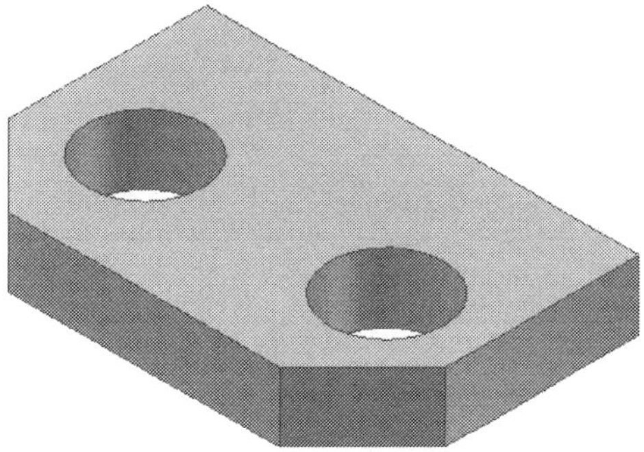

Figure 5-6 *Base feature of the model*

Creating a Feature on the Back Face of the Base Feature

To create a feature on the back face of the base feature, you first need to define the sketching plane on the back face.

1. Choose the **Start 2D Sketch** tool from **3D Model > Sketch > Start 2D Sketch** drop-down; you are prompted to select the sketching plane.

2. As the **Start 2D Sketch** tool is active, move the cursor close to the back face of the model and hover it for sometime. On doing so, the **Select Other** flyout is displayed on the model. Choose the desired face option from this flyout to select the back face of the model, refer to Figure 5-7. Next, click on the model to confirm your selection.

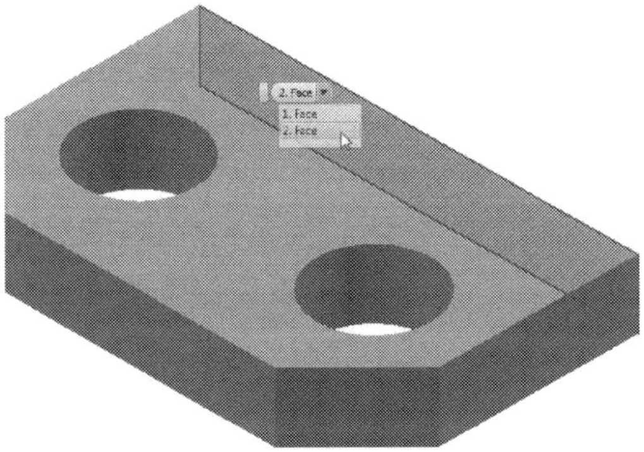

***Figure 5-7** Selecting the back face of the model using the options in the **Select Other** flyout*

3. Draw the sketch of the feature using various sketching tools, see Figure 5-8.

4. Add the required constraints to the sketch and then dimension it to make it fully constrained. The sketch after dimensioning should look similar to the one shown in Figure 5-8.

5. Choose the **Finish Sketch** button from the **Exit** panel of the **Sketch** tab and exit the Sketching environment.

Extruding the Sketch

1. Change the current view to isometric if required and then extrude the sketch up to a distance of 32 mm by using the **Extrude** tool.

   You can change the direction of the depth by choosing the **Flipped** button in the **Direction** area of the **Behavior** node and choose **OK**. The model after creating the feature on the back face will look similar to the one shown in Figure 5-9.

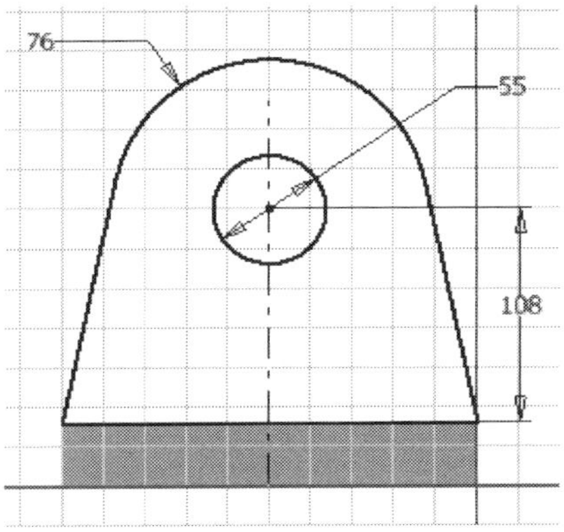

Figure 5-8 Sketch of the feature on the back face

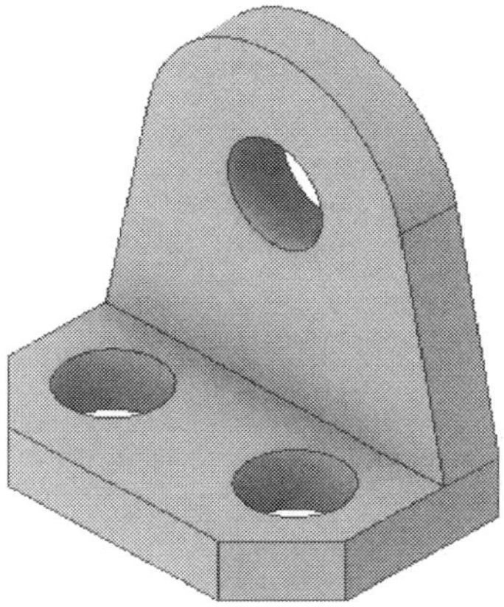

Figure 5-9 Model after creating the feature

Creating the Sketch on the Front Face of the Base Feature

1. Choose the **Start 2D Sketch** tool from the **Sketch** panel of the **3D Model** tab; you are prompted to select the plane on which the sketch will be created.

2. Select the front face of the model; the Sketching environment is invoked.

3. Delete the reference geometries, if any, and then draw a rectangle as the sketch for the next feature, as shown in Figure 5-10. Add the Collinear Constraint between the lower edge of the rectangle and the upper edge of the front face of the base feature.

4. Add required dimensions to the sketch, refer to Figure 5-10.

5. Exit the Sketching environment and then change the current view to isometric view.

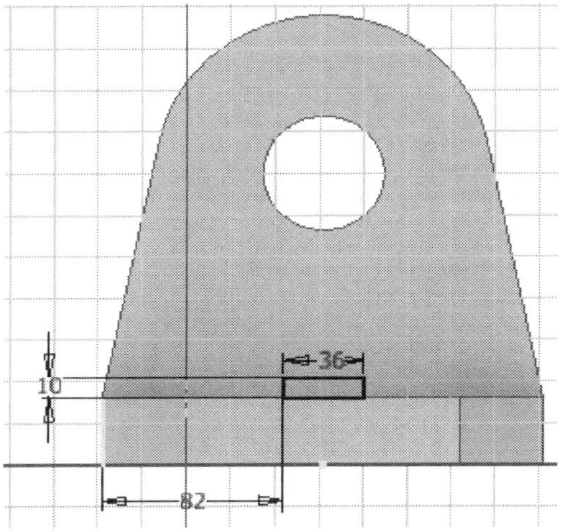

Figure 5-10 Dimensioned sketch for the feature on the front face

> **Tip**
> *Whenever you apply the Collinear constraint between a sketched line and an edge, another line will be created. It is recommended that you do not delete this line as deleting this line will delete the Collinear constraint also.*

Extruding the Sketch

1. Choose the **Extrude** tool from the Marking menu; the **Properties-Extrude** dialog box is displayed. Select the rectangle as the profile to be extruded, if it is not selected by default.

2. Choose the **To** button from the **Behavior** node; the **TO** display box is displayed and you are prompted to select the work plane or face to end the extrude.

3. Select the face, as shown in Figure 5-11, as the face where the current feature will terminate.

4. Choose the **OK** button from the dialog box. The final model for Tutorial 1 is shown in Figure 5-12.

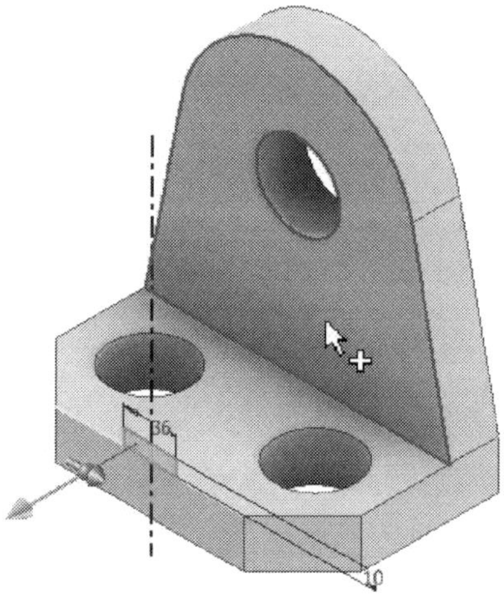

***Figure 5-11** Selecting the face to terminate the feature*

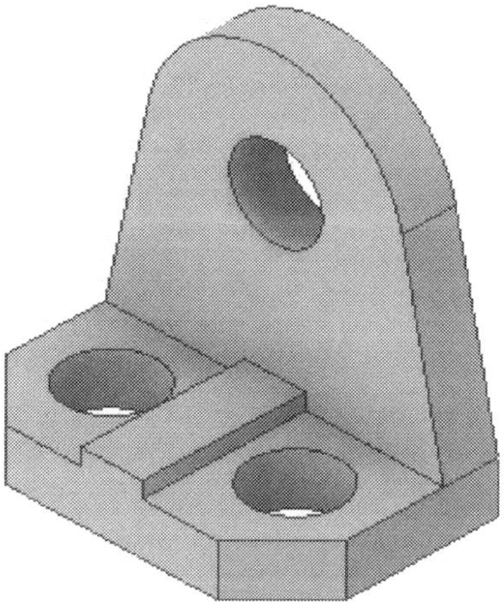

***Figure 5-12** Final model for Tutorial 1*

Saving the Model

1. Save the model with the name *Tutorial1* at the location given below and then close the file.

   C:\Inventor_2020\c05

Tutorial 2

In this tutorial, you will create the model shown in Figure 5-13. Its views and dimensions are shown in the same figure. **(Expected time: 30 min)**

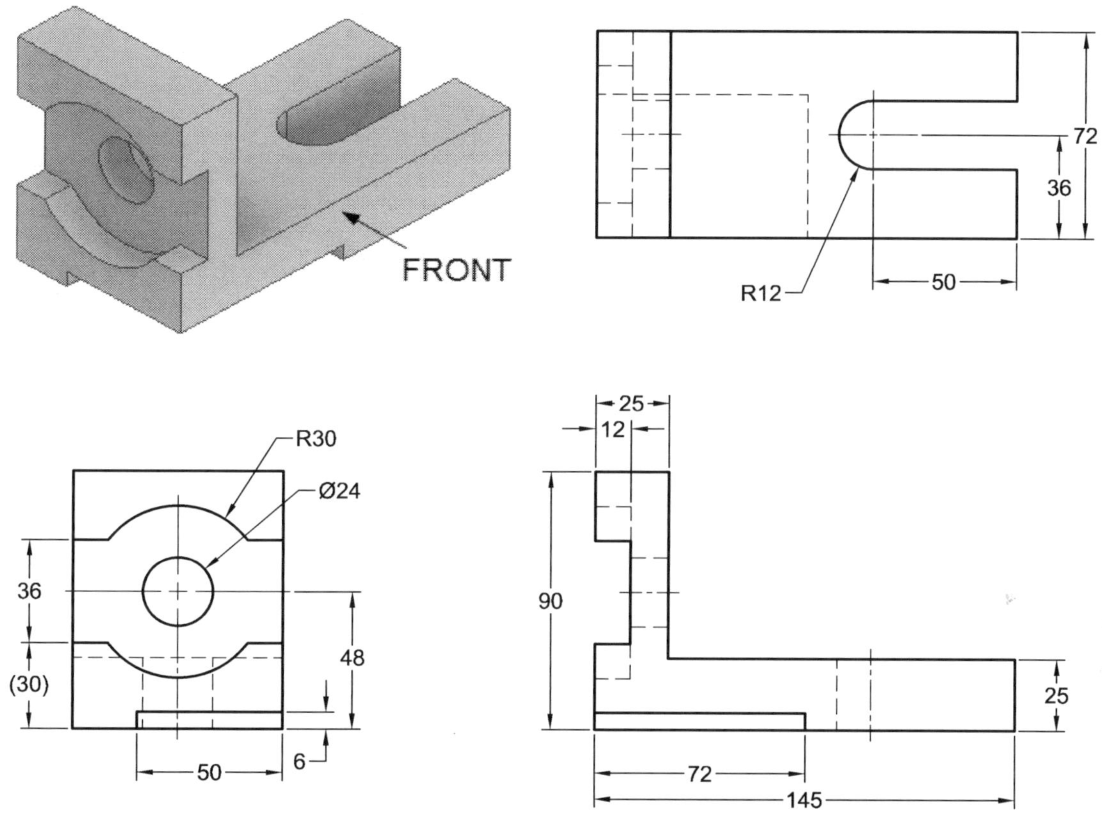

Figure 5-13 *Views and dimensions for Tutorial 2*

The following steps are required to complete this tutorial:

a. Create the base feature on the YZ plane by defining a new sketch plane on it, refer to Figures 5-14 and 5-15.
b. Define a new sketch plane on the front face of the model and create a cut feature, refer to Figures 5-16 and 5-17.
c. Create the next cut feature by defining a new sketch plane on the back face of the model, refer to Figures 5-18 and 5-19.
d. Define a new sketch plane on the new face that is exposed by creating the last cut feature and create a circular cut feature, refer to Figure 5-20.
e. Create the final cut feature on the top face of the horizontal base of the first feature, refer to Figure 5-20.

Changing the Sketch Plane

The base feature for this model is an L-shaped feature. You have to create the sketch in the YZ plane.

1. Start a new metric standard part file.

2. Choose the **Start 2D Sketch** button from the **Sketch** panel of the **3D Model** tab; the default planes are displayed and you are prompted to select the sketching plane.

3. Select the **YZ** plane as the sketching plane from the **Browser Bar**; the Sketching environment is invoked and the **YZ** plane becomes parallel to the screen. Alternatively, you can select **YZ** plane from the graphics window.

Creating and Dimensioning the Sketch of the Base Feature

1. Draw the L-shaped sketch for the base feature and add required constraints to it.

2. Add dimensions to the sketch. The sketch after adding the dimensions is shown in Figure 5-14.

3. Choose the **Finish Sketch** button from the **Exit** panel of the **Sketch** tab to exit the Sketching environment. Change the current view to the isometric view.

Extruding the Sketch

1. Choose the **Extrude** tool from the **Create** panel of the **3D Model** tab to invoke the **Properties-Extrude** dialog box. Next, extrude the sketch up to a distance of 72 mm using the **Symmetric** button. Choose **OK** to exit the **Properties-Extrude** dialog box; the base feature is created, as shown in Figure 5-15.

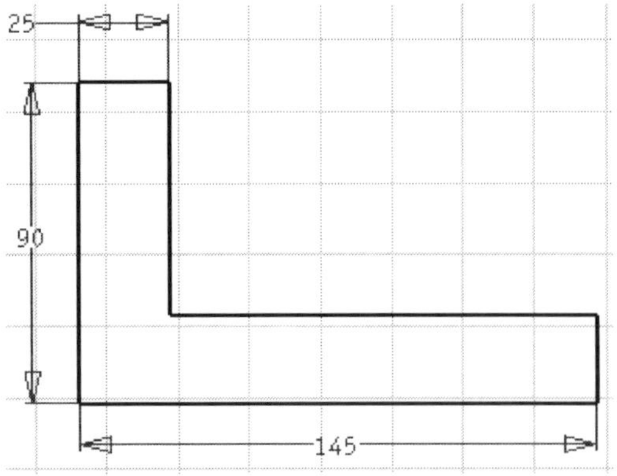

Figure 5-14 *Sketch for the base feature*

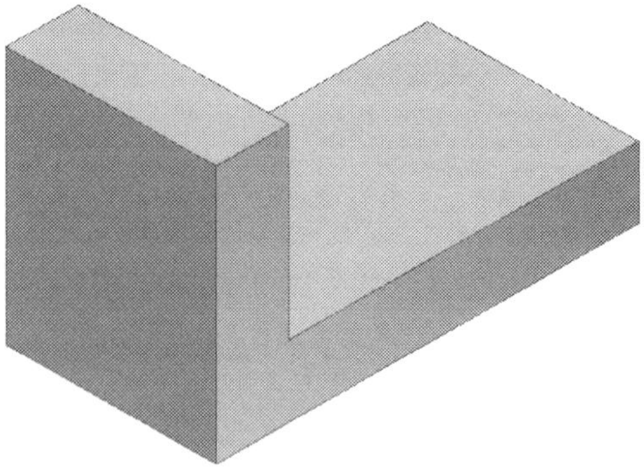

Figure 5-15 *Base feature*

Creating the Sketch for the Cut Feature on the Front Face

The next feature is a rectangular cut feature and is to be created on the front face of the base feature.

1. Choose the **Start 2D Sketch** tool from the **Sketch** panel of the **3D Model** tab; you are prompted to select the sketching plane. Select the front face of the base feature as the Sketching plane; the Sketching environment is invoked.

2. If required, reorient the model by using the ViewCube. Draw the sketch for the cut feature and delete the reference geometries if created while defining the sketch plane.

3. Next, add required constraints and dimensions to it. The dimensioned sketch is shown in Figure 5-16.

4. Choose the **Finish 2D Sketch** button from the Marking menu and then change the current view to the isometric view.

Creating the Cut Feature on the Front Face of the Model

1. Extrude the profile defined by the rectangle up to a distance of 50 mm using the **Cut** button from the **Properties-Extrude** dialog box. The isometric view of the model with the cut feature is shown in Figure 5-17.

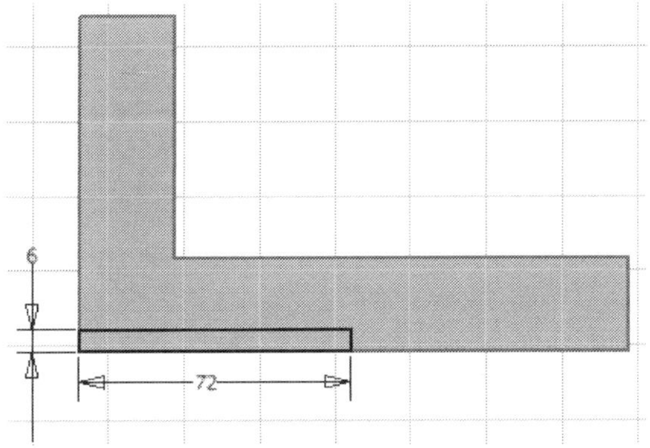

Figure 5-16 *Sketch for the cut feature*

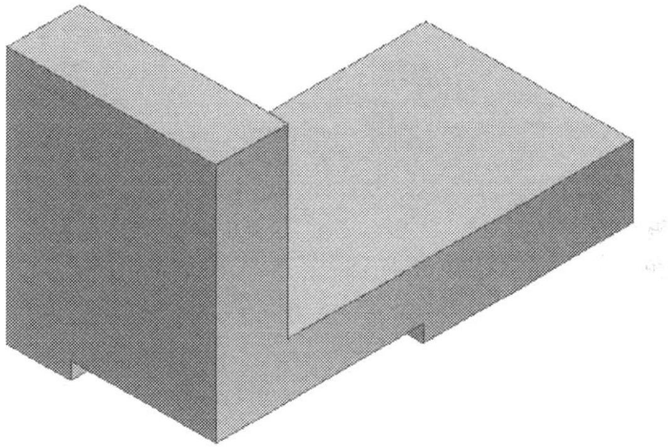

Figure 5-17 *Model after creating the cut feature on the front face*

Creating the Sketch for the Cut Feature on the Left Face

The next feature is a cut feature and is to be created on the left face of the model.

1. Choose the **Start 2D Sketch** tool from the **Sketch** panel of the **3D Model** tab; you are prompted to select a plane for creating the sketch. Select the left face of the model; the Sketching environment is activated.

2. Draw the sketch for the cut feature and delete all the reference geometries, if any. Add required constraints and dimensions to the sketch, as shown in Figure 5-18.

3. Exit the Sketching environment and then change the current view to the isometric view.

Extruding the Sketch to Create a Cut Feature

1. Extrude the profile up to a distance of 12 mm using the **Cut** operation. The model after creating a cut feature on the left face is shown in Figure 5-19.

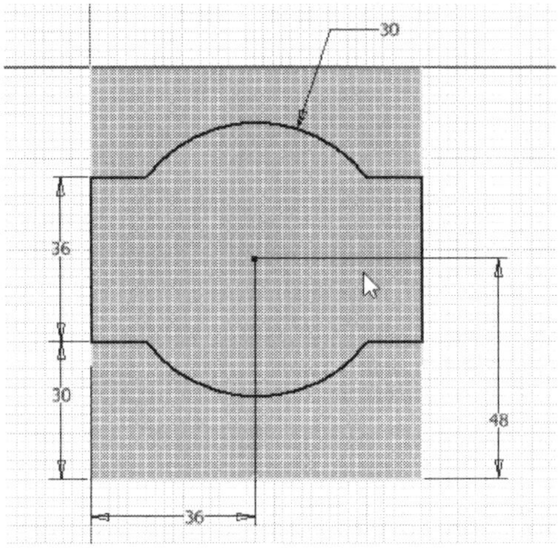

Figure 5-18 *Sketch for the cut feature*

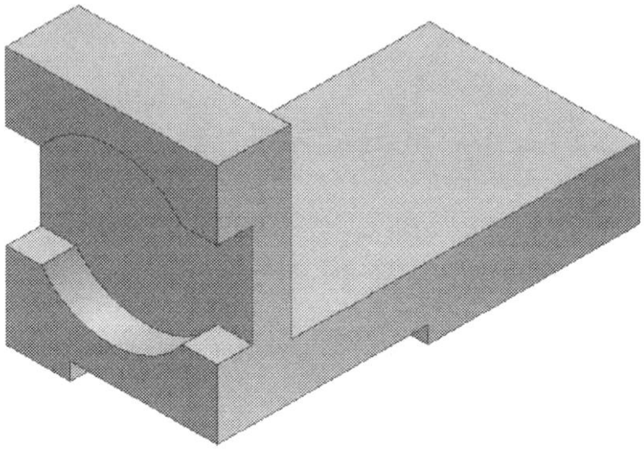

Figure 5-19 *Model after creating the cut feature on the left face*

Creating a Hole

1. Define a new sketch plane on the face that is exposed after creating the cut feature in the previous step.

2. Draw a circle on this face and then add dimensions to it, refer to Figure 5-13 for dimensions.

3. Invoke the **Extrude** tool and extrude the circle using the **Cut** operation. Note that to create this cut feature, you need to choose the **Through All** button from the **Behavio**r node next to the **Distance A** edit box. You can also choose the **Flipped** button from the **Direction** area to get the desired results.

Creating the Last Cut Feature

1. Define a sketch plane on the horizontal face of the base feature, refer to Figure 5-20, and then reorient the model using the ViewCube.

2. Delete all the reference geometries, if any, and then create the sketch for the cut feature, refer to Figure 5-13 for dimensions. Add the required constraints and dimensions to the sketch.

3. Extrude the sketch using the **Cut** operation. Use the **Through All** button from the **Behavior** node and choose **OK**. The final model for Tutorial 2 is shown in Figure 5-20.

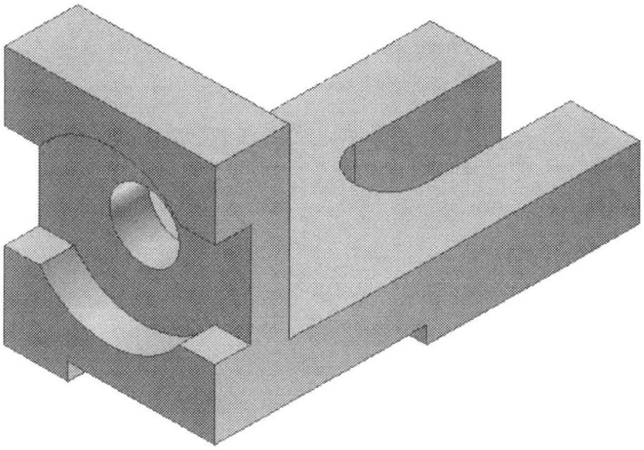

Figure 5-20 *Solid model for Tutorial 2*

Saving the Model

1. Save the sketch with the name *Tutorial2* at the location given next.

   C:\Inventor_2020\c05

2. Choose **Close > Close** from the **File** menu to close the file.

EXERCISES

Exercise 1

Create the model shown in Figure 5-21. Its dimensions are also given in the same figure.
(Expected time: 45 min)

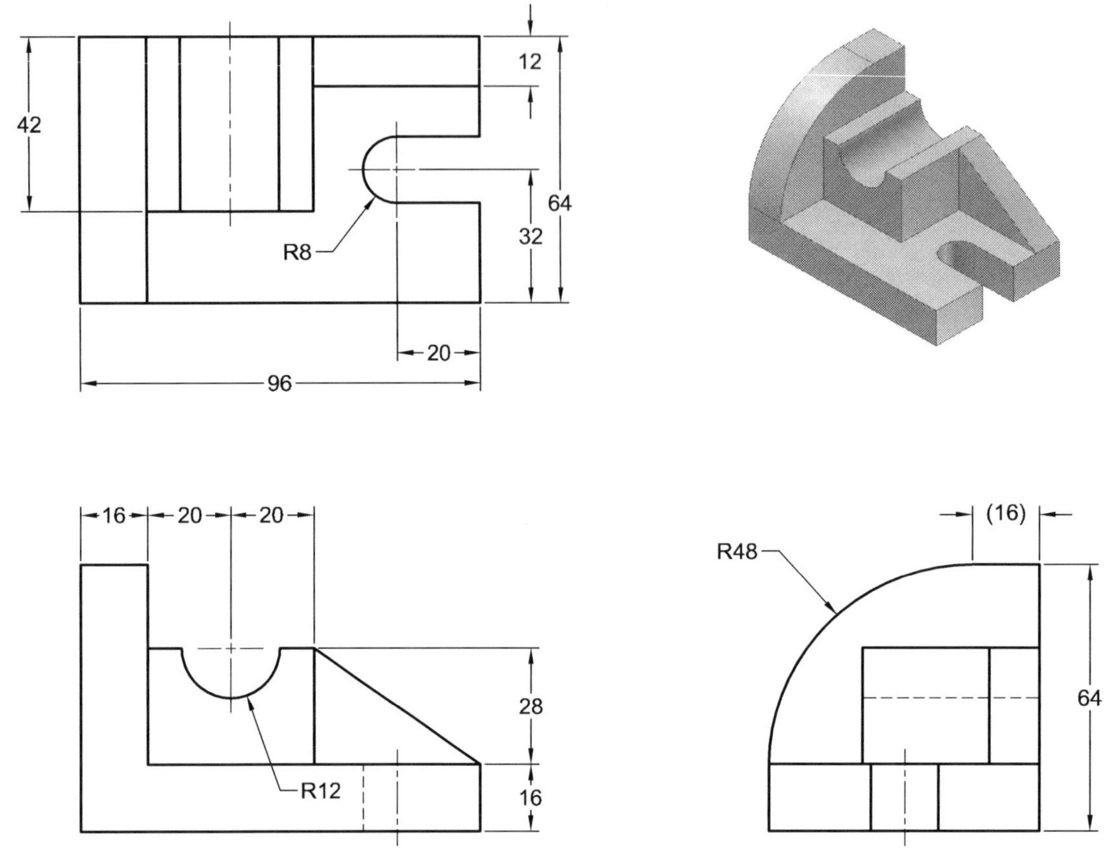

Figure 5-21 *Model and its dimensions for Exercise 1*

Exercise 2

Create the model shown in Figure 5-22. Its dimensions are given in Figure 5-23.

(Expected time: 30 min)

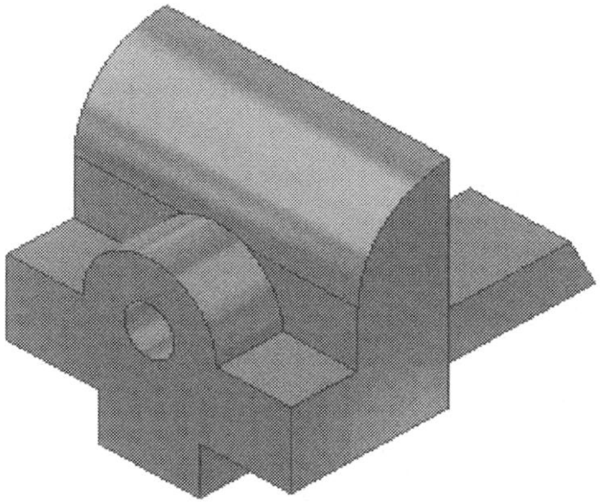

Figure 5-22 Model for Exercise 2

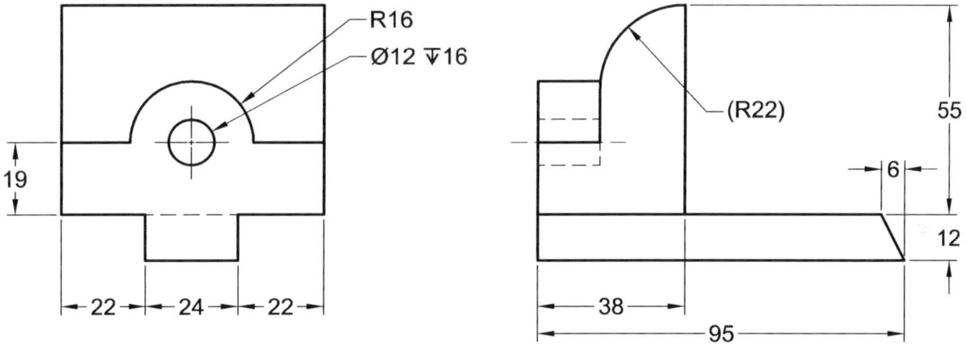

Figure 5-23 The dimensions of the model for Exercise 2

This page is intentionally left blank

Chapter 6

Advanced Modeling Tools-I

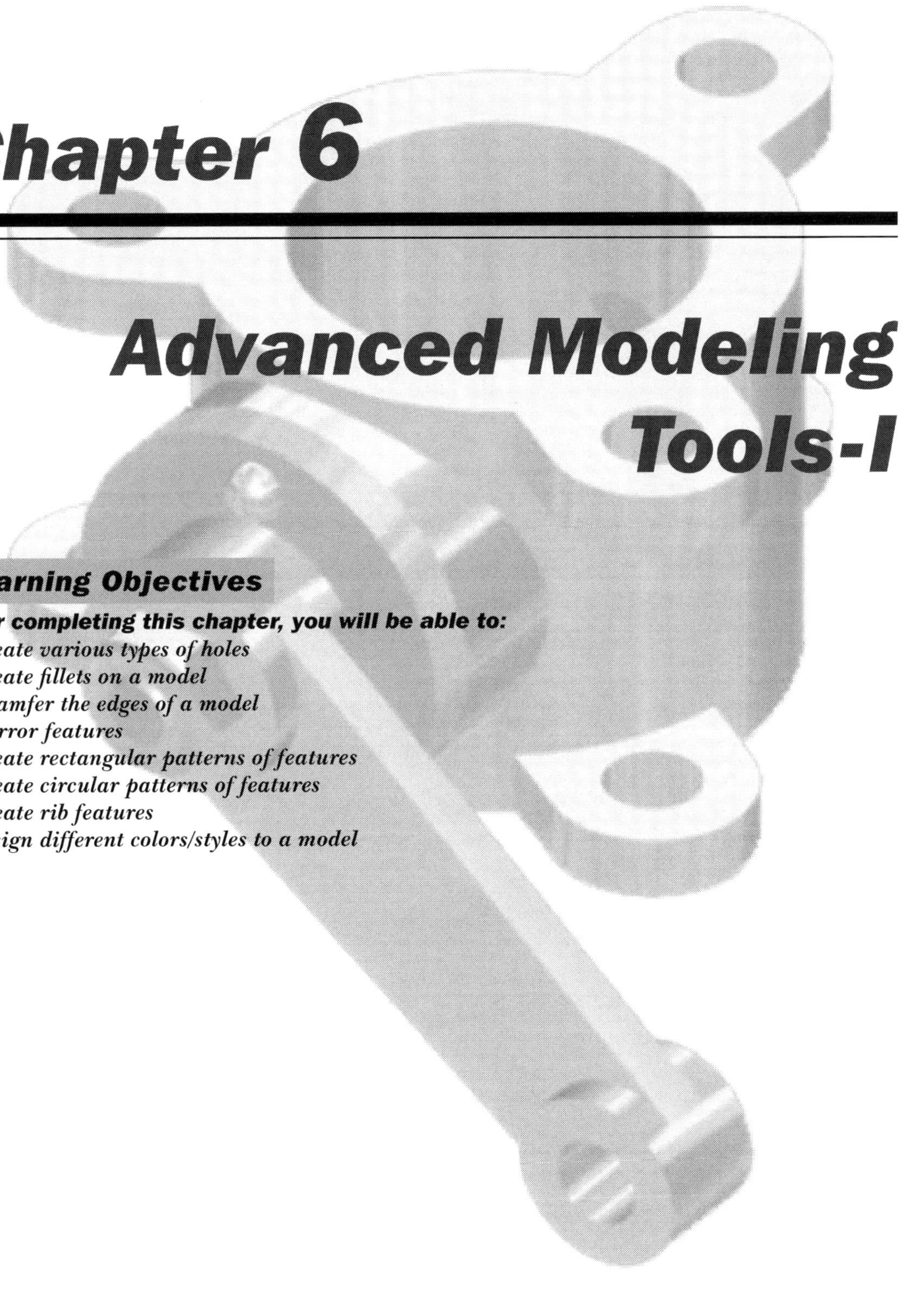

Learning Objectives

After completing this chapter, you will be able to:

- *Create various types of holes*
- *Create fillets on a model*
- *Chamfer the edges of a model*
- *Mirror features*
- *Create rectangular patterns of features*
- *Create circular patterns of features*
- *Create rib features*
- *Assign different colors/styles to a model*

ADVANCED MODELING TOOLS

Autodesk Inventor has a number of advanced modeling tools to assist you in creating a design. These advanced modeling tools appreciably reduce the time taken in creating the features in the models, thus reducing the designing time. For example, to create a hole in a cylindrical feature, one option is that while sketching the cylindrical feature, you sketch the hole also. But to edit the dimensions of the hole, you will have to edit the complete sketch. Also, if the hole is drawn along with the sketch of the cylindrical feature, it will be extruded to the same distance. However, if you want the hole to terminate before the end of the cylindrical feature, you will have to draw another sketch. But, if you create the hole using the **Hole** tool, you can specify its depth and other parameters. The advanced modeling tools used in Autodesk Inventor are listed below.

1. Hole
2. Fillet
3. Chamfer
4. Mirror
5. Rectangular Pattern
6. Circular Pattern
7. Rib
8. Thicken/Offset
9. Emboss
10. Decal
11. Sweep
12. Loft
13. Coil
14. Thread
15. Shell
16. Face Draft
17. Split
18. Boundary Patch
19. Stitch Surface
20. Replace Face
21. Delete Face
22. Move Face
23. Sculpt
24. Extend Surface

Creating Holes

Ribbon: 3D Model > Modify > Hole

Holes are circular cut features that are created on an existing feature. Holes are generally provided to accommodate fasteners in an assembly. You can create drilled, counterbore, spotface, and countersink holes using the **Hole** tool.

Creating Fillets

Ribbon: 3D Model > Modify > Fillet

In Autodesk Inventor, you can add fillets or rounds using the **Fillet** tool. Fillets are generally used to apply curves on the interior edges of a model and result in concave surfaces by adding material. Rounds are generally used to apply curves on the exterior edges and result in convex surface by removing the material.

Creating Chamfers

Ribbon: 3D Model > Modify > Chamfer

Chamfering is a process of beveling the sharp edges of a model to reduce stress concentration. In Autodesk Inventor, chamfers are created using the **Chamfer** tool.

Mirroring Features and Models

Ribbon:	3D Model > Pattern > Mirror

The **Mirror** tool is used to create the mirrored copies of selected features or to mirror the entire model by using a mirror plane. The plane that can be used to mirror the features can be a planar face or a work plane. On using this tool, an exact replica of the selected entities will be created on the other side of the mirror plane.

Creating Rectangular Patterns

Ribbon:	3D Model > Pattern > Rectangular Pattern

You can use the **Rectangular Pattern** tool to create a rectangular pattern of the selected features or surfaces, or the entire model.

Creating Circular Patterns

Ribbon:	3D Model > Pattern > Circular Pattern

In the Part module, you can use the **Circular Pattern** tool to arrange the selected features around an imaginary cylinder, thereby creating a circular pattern.

Creating Rib Features

Ribbon:	3D Model > Create > Rib

Ribs are defined as thin wall-like structures used to bind joints together so that they do not fail under an increased load. They are used to increase the stiffness of the whole structure.

Creating the Embossed and Engraved Features

Ribbon:	3D Model > Create > Emboss

The **Emboss** tool allows you to create an embossed or engraved feature. Generally, this tool is used to emboss or engrave text on an existing feature. This tool remains inactive until a sketch or a text is available in the graphics window.

TUTORIALS

Tutorial 1

In this tutorial, you will create the model of a Fixture Base shown in Figure 6-1. Its dimensions are given in the same figure. After creating the solid model, you will change its color to yellow.

(Expected time: 45 min)

The following steps are required to complete this tutorial:

a. Start a new part file and invoke the Sketching environment. Create the sketch for the base feature on the XY plane and extrude it up to a distance of 102 mm.

b. Define a new sketch plane on the back face of the base feature and create the join feature.
c. Create two cylindrical features with holes on the front face of the second feature1.
d. Create the fillet on the base feature.
e. Create two counterbore holes taking the reference of the cylindrical faces of fillets by using the **Hole** tool.
f. Finally, draw an open sketch and convert it into a rib using the **Rib** tool to complete the model.
g. Change the appearance of the model by using the **Appearance** drop-down list in the **Quick Access Toolbar**.

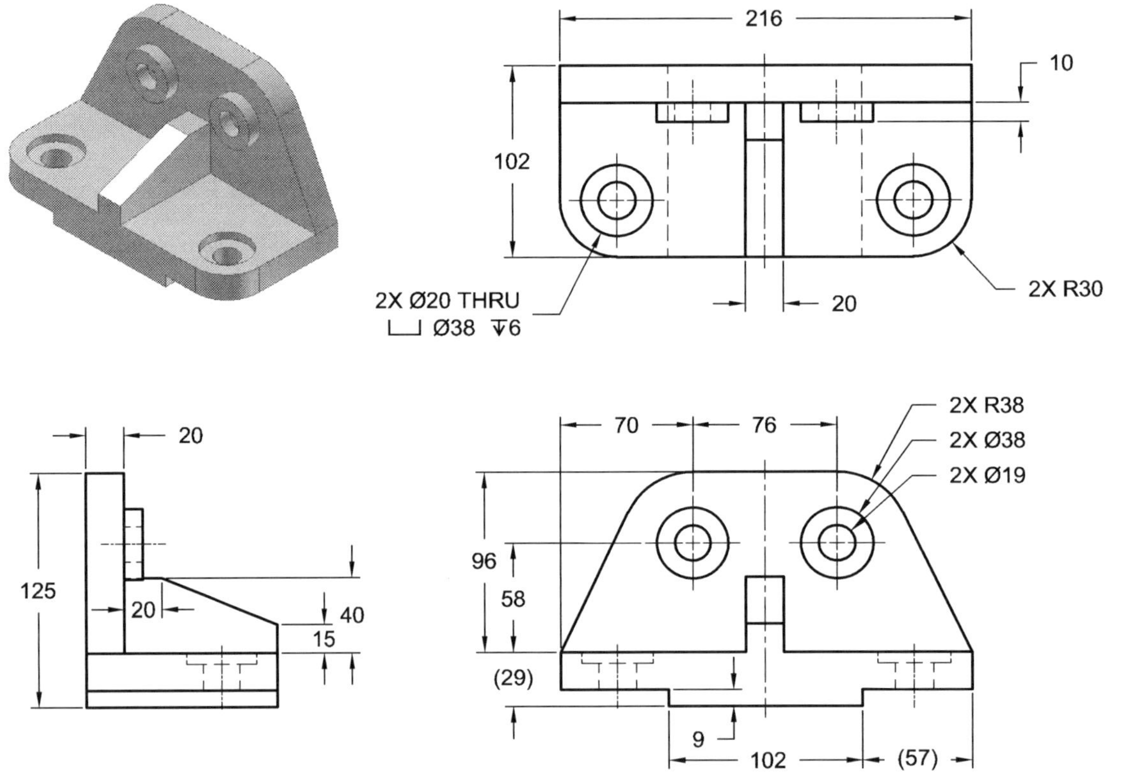

Figure 6-1 *Views and dimensions of the model for Tutorial 1*

Creating the Base Feature

You need to create the base feature on the XY plane.

1. Start a new metric standard part file.

2. Choose the **Start 2D Sketch** button from the **Sketch** panel of the **3D Model** tab; the default planes are displayed and you are prompted to select the sketching plane.

3. Select the **XY** plane as the sketching plane from the **Browser Bar**; the Sketching environment is invoked and the **XY** plane becomes parallel to the screen. Alternatively, select the **XY** plane from the graphics window.

4. Create sketch for the base feature. Add the required constraints and dimensions to it. The sketch after adding constraints and dimensions is shown in Figure 6-2.

5. Exit the Sketching environment and extrude the sketch up to a distance of 102 mm using the **Extrude** tool to create the base feature. The base feature is shown in Figure 6-3.

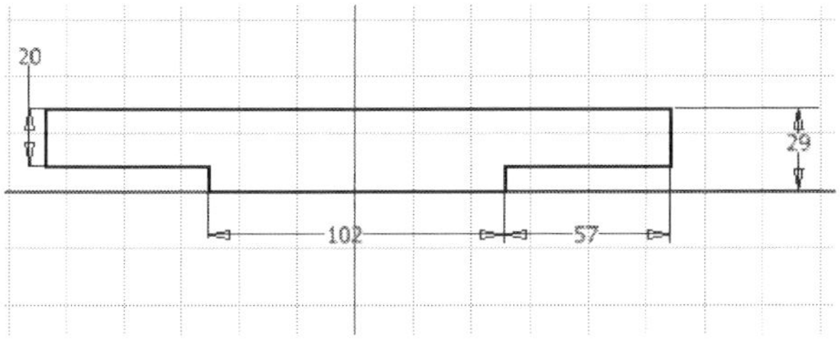

Figure 6-2 *Sketch for the base feature*

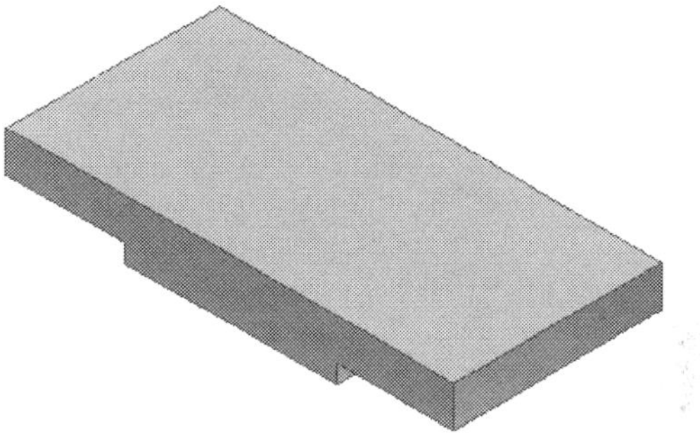

Figure 6-3 *Base feature*

Creating a Join Feature on the Back Face of the Base Feature

1. Define a new sketch plane on the back face of the base feature. Draw the sketch for the join feature and then add required constraints and dimensions to it. The sketch after adding constraints and dimensions is shown in Figure 6-4.

2. Exit the Sketching environment and then extrude the sketch up to a distance of 20 mm toward the front of the base feature, as shown in Figure 6-5.

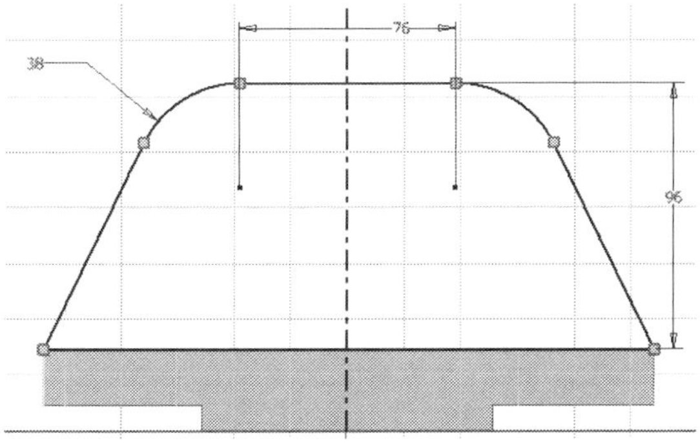

Figure 6-4 *Sketch for the join feature*

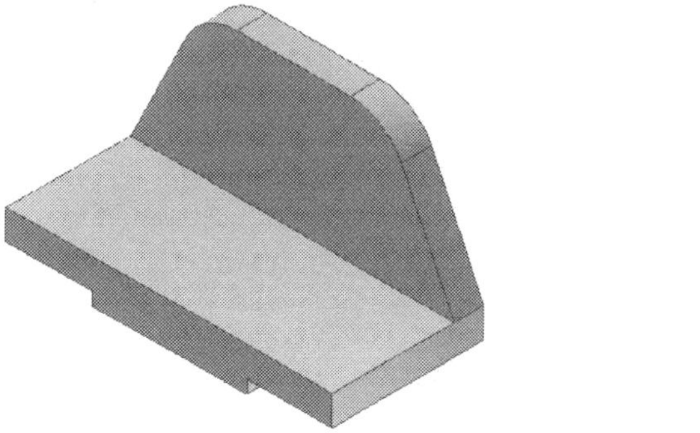

Figure 6-5 *Model after creating the join feature*

Creating Cylindrical Features on the Front Face of the Second Feature

To create two cylindrical features, you need to draw a sketch consisting of two concentric circles. The reason for drawing the sketch for both the features together is that both the cylindrical features are to be extruded to the same distance.

1. Define a new sketch plane on the front face of the second feature and draw the sketches for both the cylindrical features, as shown in Figure 6-6.

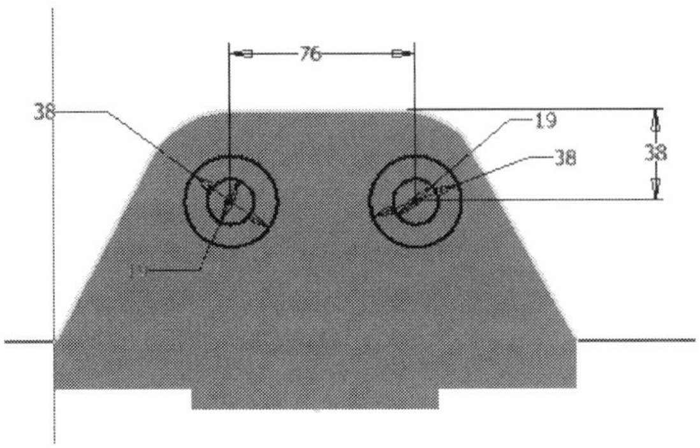

Figure 6-6 *Sketches for the cylindrical features*

2. Invoke the **Extrude** tool and extrude the sketches to a distance of 10 mm.

   While selecting profiles for extrusion in both sketches, make sure that you click between the inner and outer circles. As a result, the inner circles are subtracted from the outer circles when you extrude the sketch, thus creating holes. The model after creating the cylindrical features is shown in Figure 6-7.

Figure 6-7 *Model after creating cylindrical features*

Creating Fillets

The vertical edges of the front face of the base feature need to be filleted so that you can use the cylindrical faces of fillets to define the center of the counterbore holes.

1. Choose the **Fillet** tool from the **Modify** panel of the **3D Model** tab; the **Fillet** dialog box is displayed and you are prompted to select the edges to be filleted. By default, the **Constant** tab is chosen in this dialog box.

2. Select the outer left and outer right vertical edges on the front face of the base feature. On selecting the edges, the **Edges** column displays **2 selected** edge and a preview of the fillet is displayed on the model with 2 mm as the default radius value.

3. Click on the default radius value in the **Radius** column and enter **30** in the edit box displayed. Alternatively, enter **30** in the edit box of the mini toolbar. You will notice that the fillet in the preview of the model has also increased accordingly. Choose the **OK** button to exit the **Fillet** dialog box; the fillets are created, as shown in Figure 6-8.

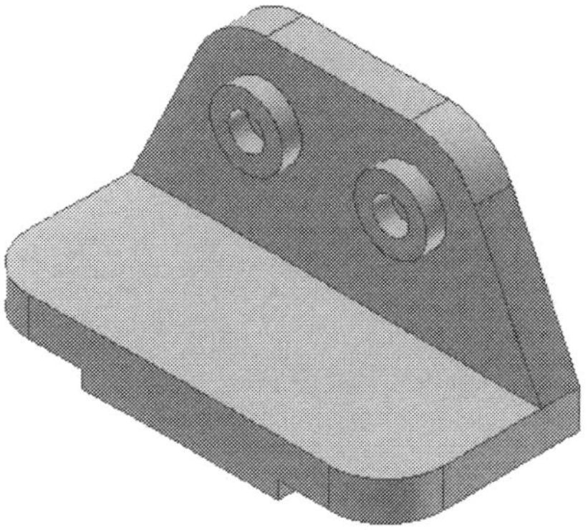

Figure 6-8 *Model after creating fillets*

Creating Counterbore Holes

As mentioned earlier, in Autodesk Inventor, you can create holes concentric to the cylindrical faces. To create two counterbore holes, you need to use the cylindrical faces of the fillet.

1. Choose the **Hole** tool from the **Modify** panel of the **3D Model** tab to invoke the **Hole** dialog box.

2. Choose the **Counterbore** button from the **Type** node.

3. Select the top planar face of the base feature as the face to place the hole; a preview of the counterbore hole with the current values is displayed.

4. Select the cylindrical face of the fillet on the right; the preview of the hole is relocated.

5. Choose the **Through All** button from the **Termination** area. Modify the value of the counterbore diameter in the preview window to **38**. Similarly, modify the value of the bore diameter to **20** and the counterbore depth to **6**. Choose **OK** to close the **Hole** dialog box.

6. Similarly, using the options already set in the **Hole** dialog box, create another hole concentric to the fillet on the left.

7. The model after creating the counterbore holes is shown in Figure 6-9.

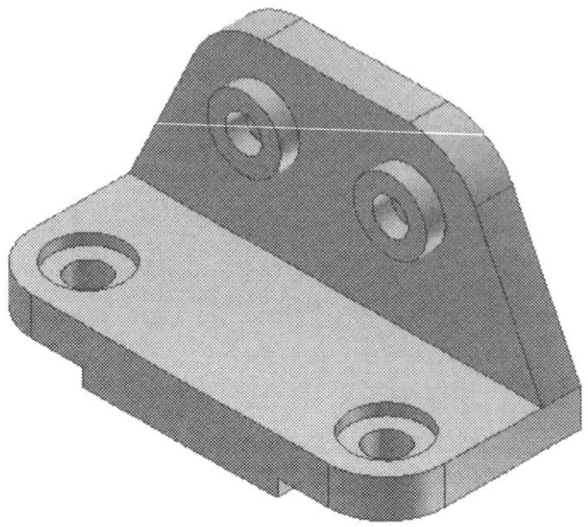

Figure 6-9 *Model after creating counterbore holes*

Creating the Rib Feature

The rib feature is created at the center of the model. Therefore, you need to define an offset work plane at the center on which the rib feature will be created.

1. Choose the **Offset from Plane** tool from **3D Model > Work Features > Plane** drop-down and select the right face of the base feature; a preview of the work plane along with the mini toolbar is displayed in the graphics window.

2. Enter **-108** in the edit box available in the mini toolbar and then choose **OK** from it. Negative value ensures that the work plane gets created inside the model. Select this work plane as the sketching plane.

3. Draw an open sketch for the rib feature and then add required constraints and dimensions to it, as shown in Figure 6-10.

   When you apply the **Coincident** constraint between the lines in the sketch and the edges of the model, the lines defining the edges are drawn. Make sure these lines are not selected when you select the sketch for creating the rib feature.

4. Exit the Sketching environment. Next, invoke the **Rib** dialog box and choose the **Parallel to Sketch Plane** button from the **Rib** dialog box. Select the open profile. Make sure that the **Direction 1** button is chosen in the **Shape** area.

5. Set the value in the edit box in the **Thickness** area to **20**. Choose **OK** to exit the **Rib** dialog box. The final model after creating all features is shown in Figure 6-11.

Changing the Appearance of the Model

When a model is created, the default color is applied to it. However, in Autodesk Inventor, you can change the default color/style of the model.

1. Select the **Yellow** option from the **Appearance** drop-down list on the right of the **Quick Access Toolbar**; the color of the model changes to yellow. Note that you do not need to select the model to apply color to it.

2. Save the model with the name *Tutorial1* at the location *C:\Inventor_2020\c06* and then close the file.

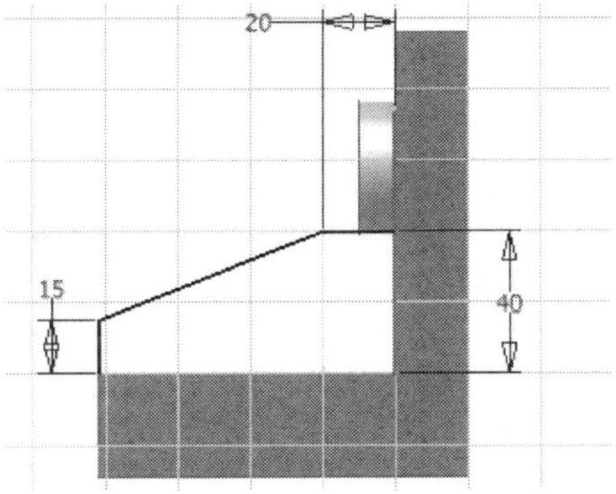

Figure 6-10 *Sketch for the rib feature*

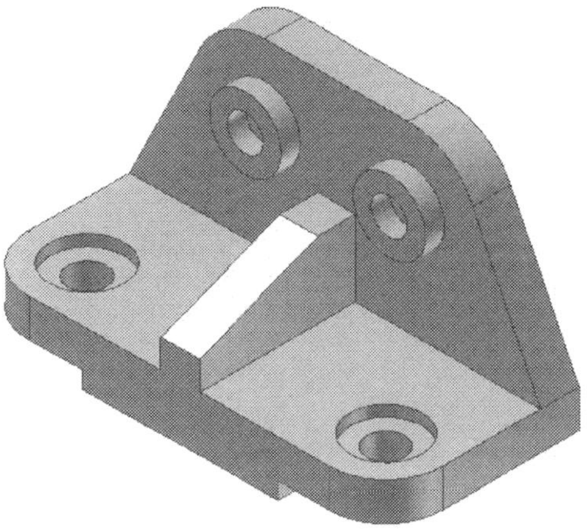

Figure 6-11 *Final model for Tutorial 1*

Tutorial 2

In this tutorial, you will create a model of the Pivot Base shown in Figure 6-12. Its dimensions are given in the same figure. Change the color/style of the model to Zinc Chromate. **(Expected time: 45 min)**

The following steps are required to complete this tutorial:

a. Create the base feature on the XY plane.
b. Create the join feature on the back face of the base feature.
c. Create another join feature on the front face of the second feature.
d. Create the cut feature on the third feature.
e. Create the rib and the join feature on the right side of the model.
f. Mirror the rib and the join feature on the left side of the model.
g. Create a hole on the top face of the base feature.
h. Change the style of the model by using the **Appearance** drop-down list in the **Quick Access Toolbar**.

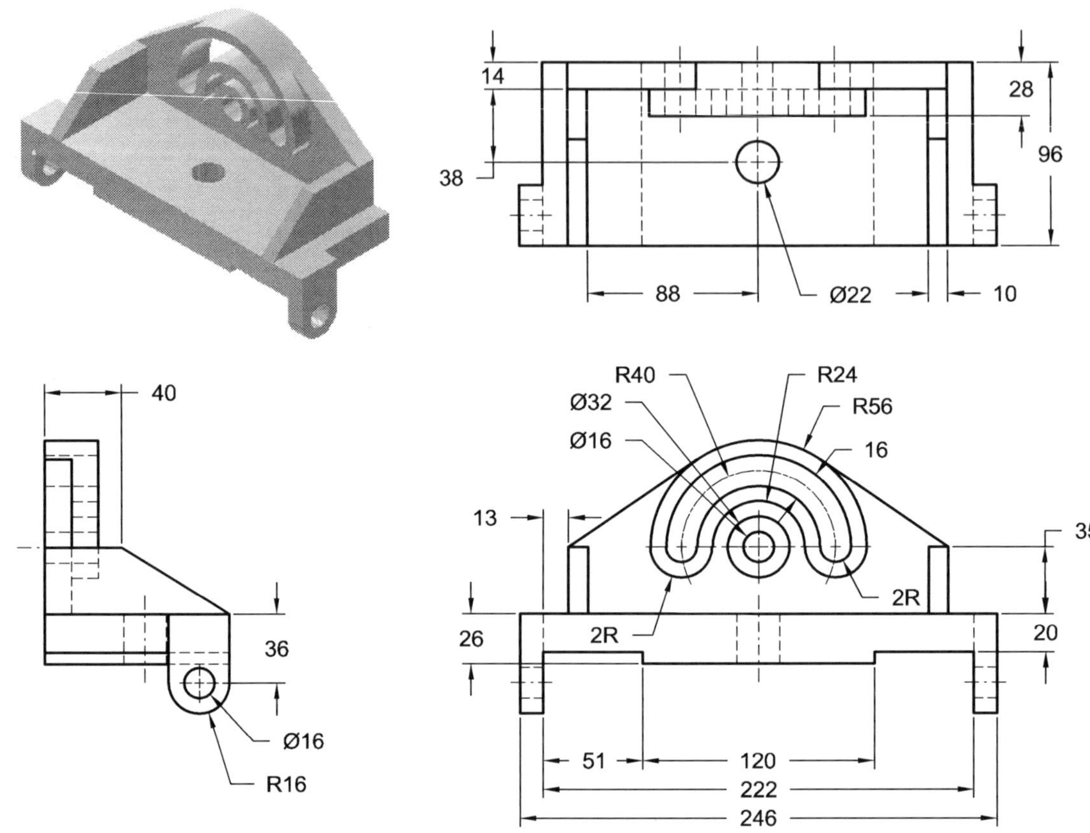

Figure 6-12 Views and dimensions of the model for Tutorial 2

Creating the Base Feature

1. Start a new metric part file.

2. Choose the **Start 2D Sketch** button from the **Sketch** panel of the **3D Model** tab; the default planes are displayed and you are prompted to select the sketching plane.

3. Select the **XY** plane as the sketching plane from the **Browser Bar**; the Sketching environment is invoked and the **XY** plane becomes parallel to the screen. Alternatively, select the **XY** plane from the Graphics window.

4. Draw the sketch of the base feature, as shown in Figure 6-13.

5. Exit the Sketching environment.

6. Click on the sketch in the graphics window; a mini toolbar with the **Create Extrude**, **Create Revolve**, and **Edit Sketch** tools is displayed.

7. Choose the **Create Extrude** tool; the **Extrude** dialog box is displayed. Also, a preview of the extrude feature along with the mini toolbar is displayed.

8. Enter **96** in the edit box of the mini toolbar and choose **OK** to create the base feature of the model, as shown in Figure 6-14.

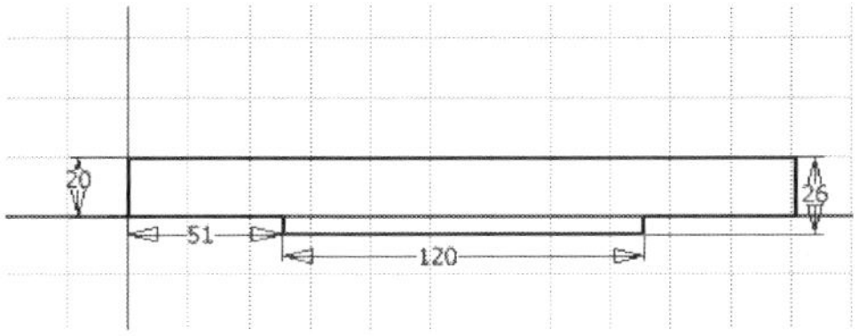

Figure 6-13 *Sketch of the base feature*

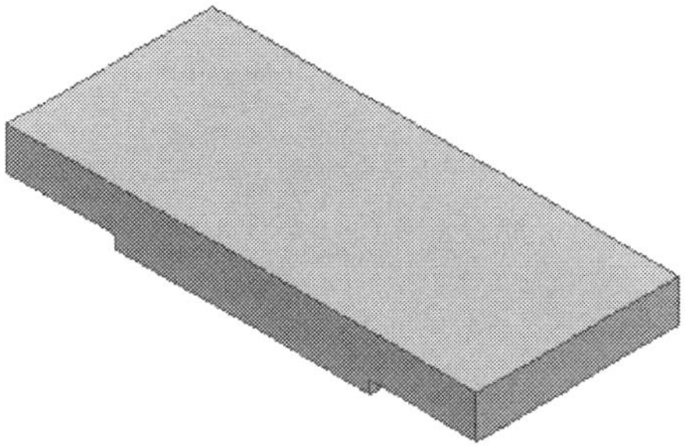

Figure 6-14 *Base feature*

Creating a Join Feature on the Back Face of the Base Feature

1. Rotate the model by using the ViewCube such that its back face is visible.

2. Select the back face of the base feature.

3. Choose the **Start 2D Sketch** tool from the **Sketch** panel and draw the sketch for the join feature on it, as shown in Figure 6-15.

4. Extrude the sketch up to a distance of 14 mm toward the front of the base feature by using the **Extrude** tool, see Figure 6-16.

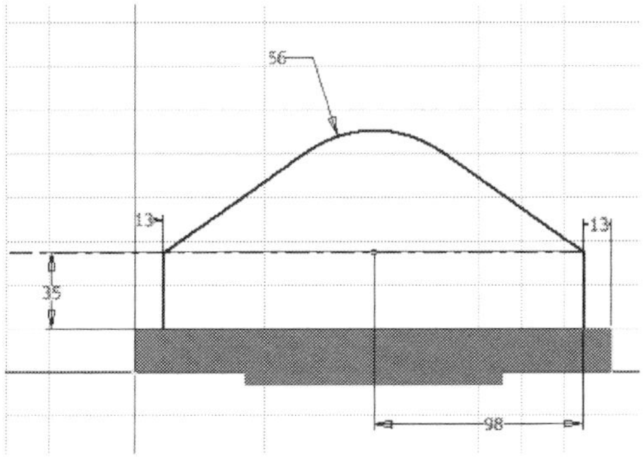

Figure 6-15 *Sketch of the join feature*

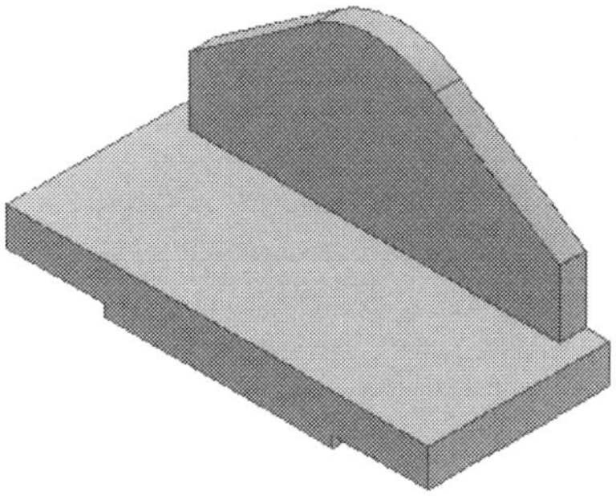

Figure 6-16 *Model after creating the join feature*

Creating the Join Feature on the Front Face of the Second Feature

1. Define a new sketch plane on the front face of the second feature.

2. Draw two disjoint sketches of the join feature, as shown in Figure 6-17.

3. Extrude both the sketches up to a distance of 14 mm using the **Extrude** tool. The model after creating the join feature is shown in Figure 6-18. Note that the two sketches are displayed as a single feature in the **Browser Bar** because both the sketches are extruded together.

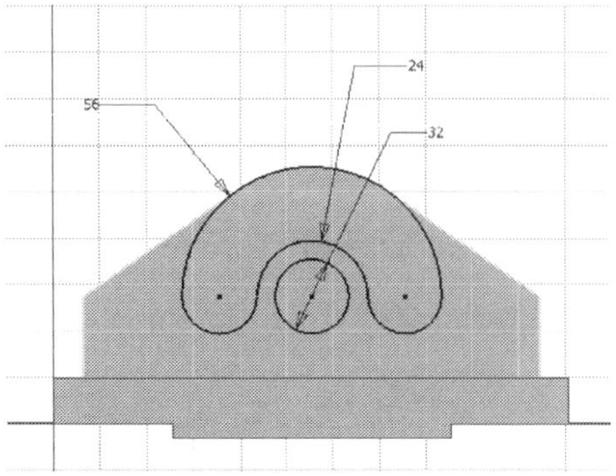

Figure 6-17 *Sketches of the join feature on the front face*

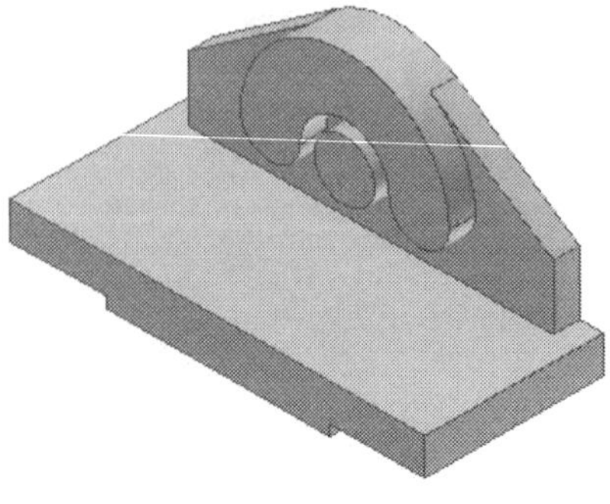

Figure 6-18 *Join feature on the front face of the second feature*

Creating the Cut Feature

Next, you need to create a cut feature that will remove material from the previous feature. You need to draw two disjoint sketches for the cut feature at the same time and then extrude them using the **Cut** operation. As the sketches are to be extruded through the model, you need to select both of them together while selecting the profile for creating the cut feature.

1. Define a new sketch plane on the front face of the semicircular feature. Then, draw the sketch for the cut feature and the circle for the hole, refer to Figure 6-19.

2. Next, invoke the **Properties-Extrude** dialog box and select the sketch. Now, choose the **Cut** button from the **Boolean** area and the **Through All** button from the **Behavior** node to create the cut feature, shown in Figure 6-20.

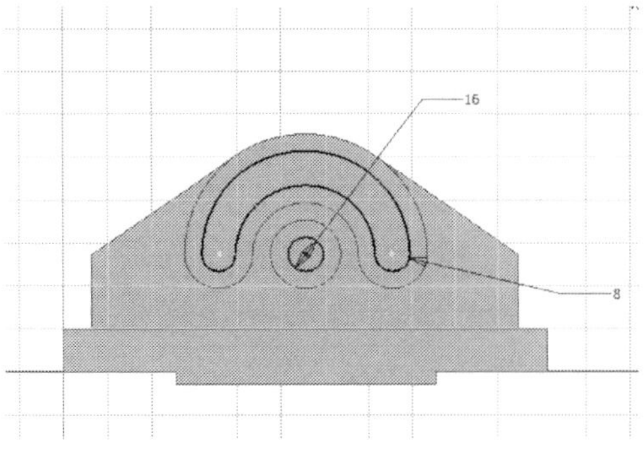

Figure 6-19 *Sketch for the cut feature*

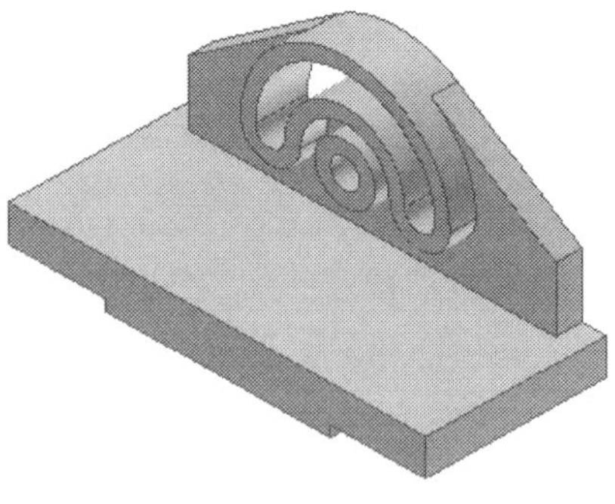

Figure 6-20 *Model after creating the cut feature*

Creating the Rib Feature

The sketch for the rib will be created on the sketch plane defined on the right face of the second feature. The sketch will be extruded toward the left to create the feature.

1. Define a new sketch plane on the right face of the second feature and then draw the sketch of the rib feature. Add the required dimensions and constraints to the sketch, as shown in Figure 6-21.

2. Exit the Sketching environment and choose the **Rib** tool from the **Create** panel of the **3D Model** tab; the **Rib** dialog box is displayed. By default, the **Normal to Sketch Plane** button is chosen in the Type Specification area. Choose the **Parallel to Sketch Plane** button in this area.

   When you apply the **Coincident** constraint between the lines in the sketch and the edges of the model, the lines defining the edges are drawn. Make sure these lines are not selected when you select the sketch for creating the rib feature.

3. Select the open sketch. Choose the second direction button in the **Thickness** area to extrude the feature toward left. Next, choose the **Direction1** button from the dialog box; a preview of the rib feature is displayed.

4. Enter **10** in the **Thickness** edit box in the **Thickness** area.

5. Choose **OK** to exit the **Rib** dialog box; the rib is created, as shown in Figure 6-22.

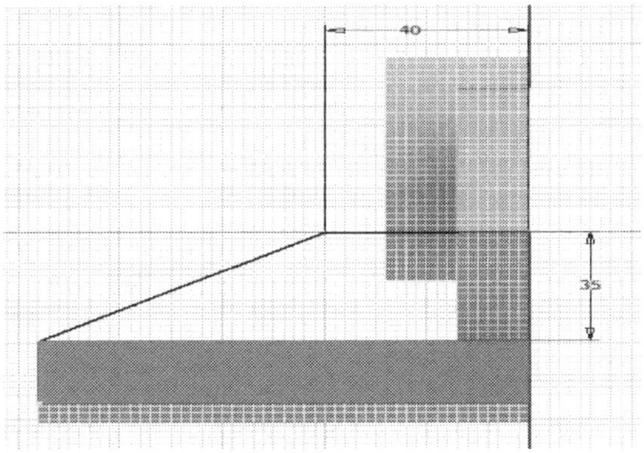

Figure 6-21 *Sketch of the rib*

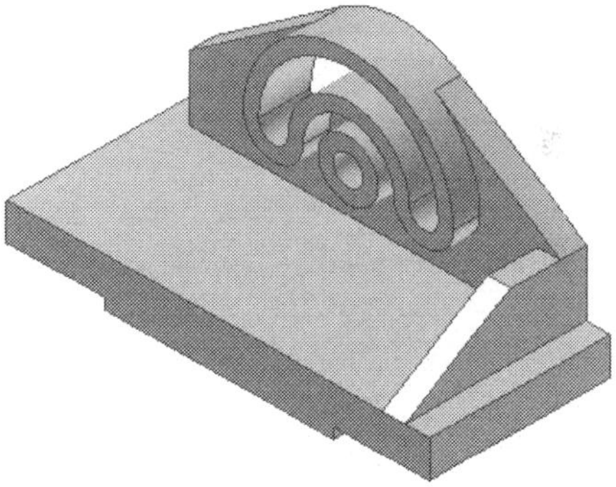

Figure 6-22 *Model after creating the rib*

Creating the Join Feature on the Right Face of the Base Feature

1. Define a new sketch plane on the right face of the base feature.

2. Draw the sketch of the join feature. Draw a circle inside the sketch such that when extruded, a hole is created automatically. Add the required constraints and dimensions to it, as shown in Figure 6-23, and then exit the Sketching environment.

3. Extrude the sketch up to a distance of 12 mm to create the feature. Make sure that you select the profile by using a point inside the outer loop but outside the circle. The model after creating the join feature is shown in Figure 6-24.

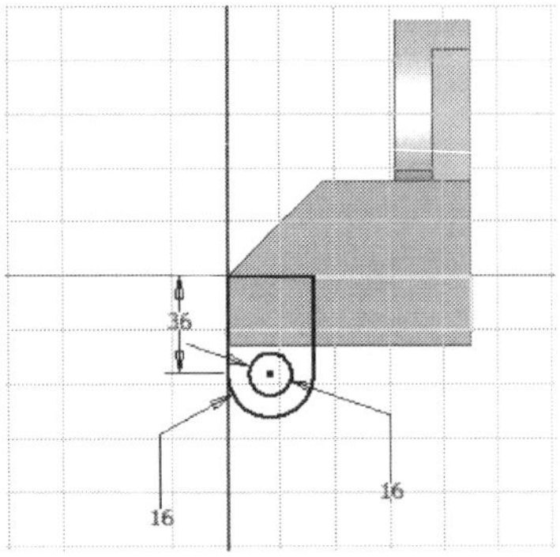

Figure 6-23 Sketch of the join feature

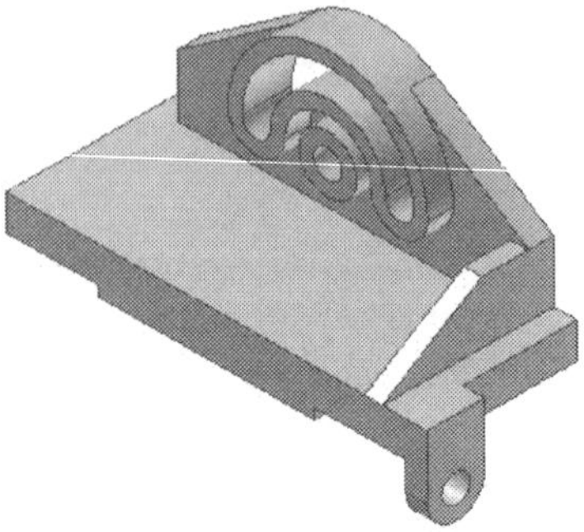

Figure 6-24 Model after creating the join feature

Mirroring Features on the other Side of the Model

The second set of rib and join features will be created by mirroring the first set of these features on the other side of the model. The features will be mirrored about the offset work plane created at the center of the model.

1. Choose the **Offset From Plane** tool from **3D Model > Work Features > Plane** drop-down; you are prompted to select a planar surface.

2. Select the right face of the base feature; a preview of the plane along with the mini toolbar is displayed.

3. Enter **-111** in the edit box of the mini toolbar and then choose **OK**; the plane is created.

4. Choose the **Mirror** tool from the **Pattern** panel of the **3D Model** tab; the **Mirror** dialog box is displayed and you are prompted to select the features to be patterned. Select the rib feature and the feature created on the right face of the base feature.

5. Choose the **Mirror Plane** button and select the offset work plane as the mirror plane; a preview of the mirrored features is displayed. Choose **OK** to exit this dialog box.

6. Right-click on the work plane in the **Browser Bar** to display the shortcut menu. Choose the **Visibility** option from the shortcut menu to turn off the visibility of the work plane. The model after mirroring the features is shown in Figure 6-25.

Creating the Hole on the Top Face of the Base Feature

1. Choose the **Hole** tool from the **Modify** panel of the **3D Model** tab to invoke the **Hole** dialog box.

2. Choose the **Simple Hole** button from the **Type** area.

3. Select the top planar face of the base feature. Next, select the edge labeled 1 on the top face of the base feature, refer to Figure 6-26; the mini toolbar is displayed.

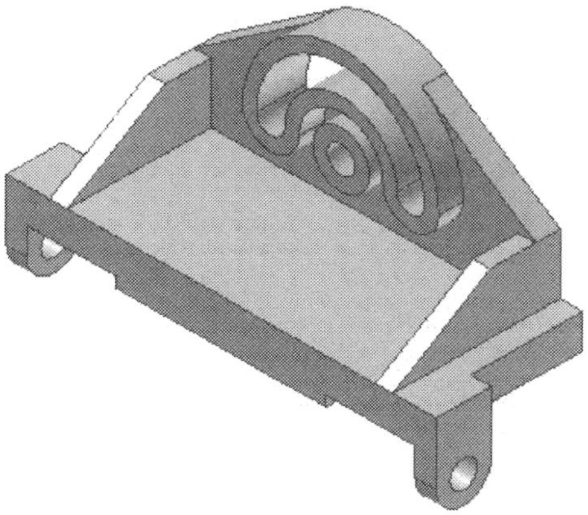

Figure 6-25 Model after mirroring the features

4. Modify the value in the edit box of the mini toolbar to **88**.

5. Similarly, select the edge labeled 2 on the top face of the base feature and modify the value to **38**, refer to Figure 6-26.

6. Choose the **Through All** button from the **Termination** area.

7. Set the value of the diameter of the hole in the preview window to **22**; the diameter of the hole in the preview also increases automatically. Choose the **OK** button. The final model for Tutorial 2 is shown in Figure 6-27.

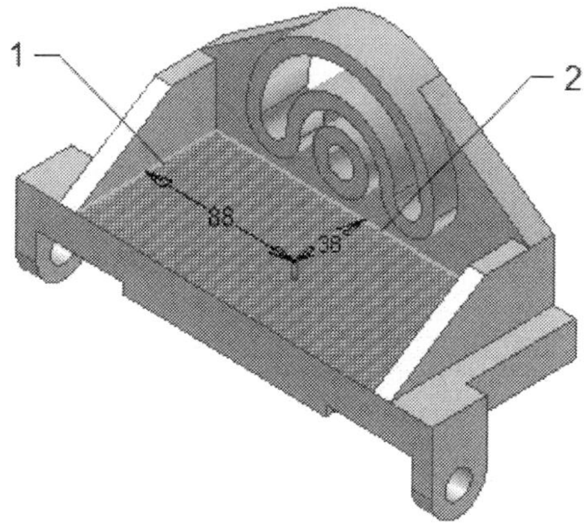

Figure 6-26 The preview of the hole feature

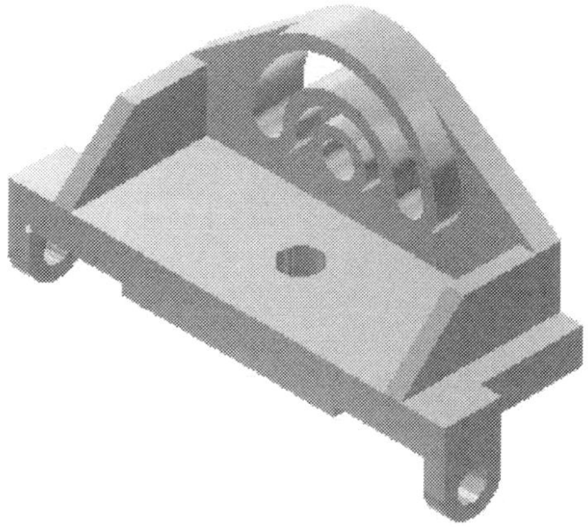

Figure 6-27 Final model for Tutorial 2

Changing the Style and Saving the Model

As mentioned earlier, the style of the feature is changed using the drop-down list next to the **Appearance** option in the **Quick Access Toolbar**.

1. Select the **Zinc Chromate 2** option from the drop-down list next to the **Appearance** option available on the extreme right of the **Quick Access Toolbar**; the material of the model is changed to Zinc Chromate.

2. Save the model with the name *Tutorial2* at the location *C:\Inventor_2020\c06* and then close the file.

EXERCISES

Exercise 1

Create the model shown in Figure 6-28. Its dimensions are shown in Figure 6-29.

(Expected time: 45 min)

Figure 6-28 *Model for Exercise 1*

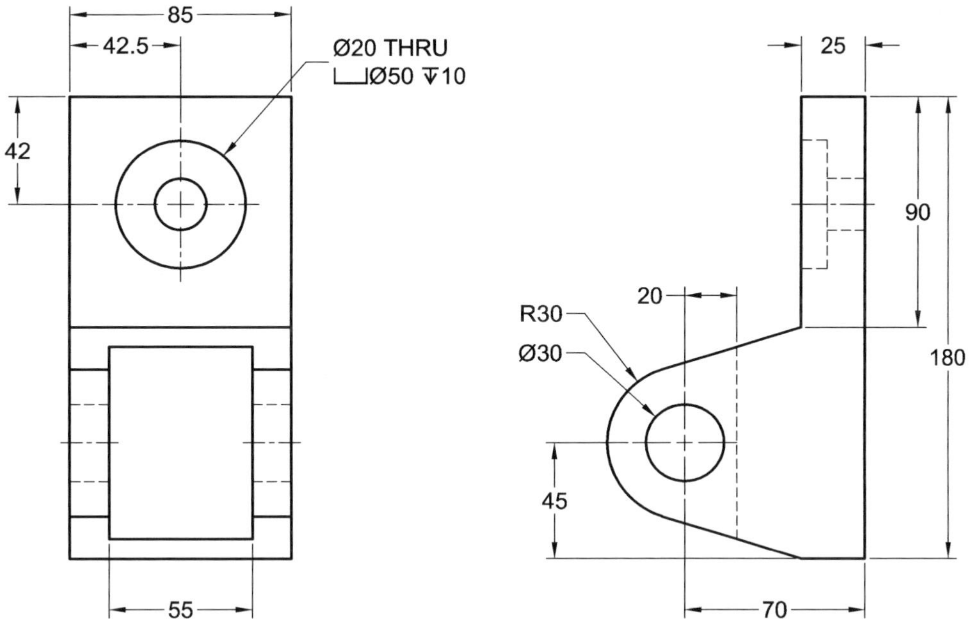

Figure 6-29 *Views and dimensions of the model*

Exercise 2

Create the model shown in Figure 6-30. Its dimensions are shown in the same figure.

(Expected time: 45 min)

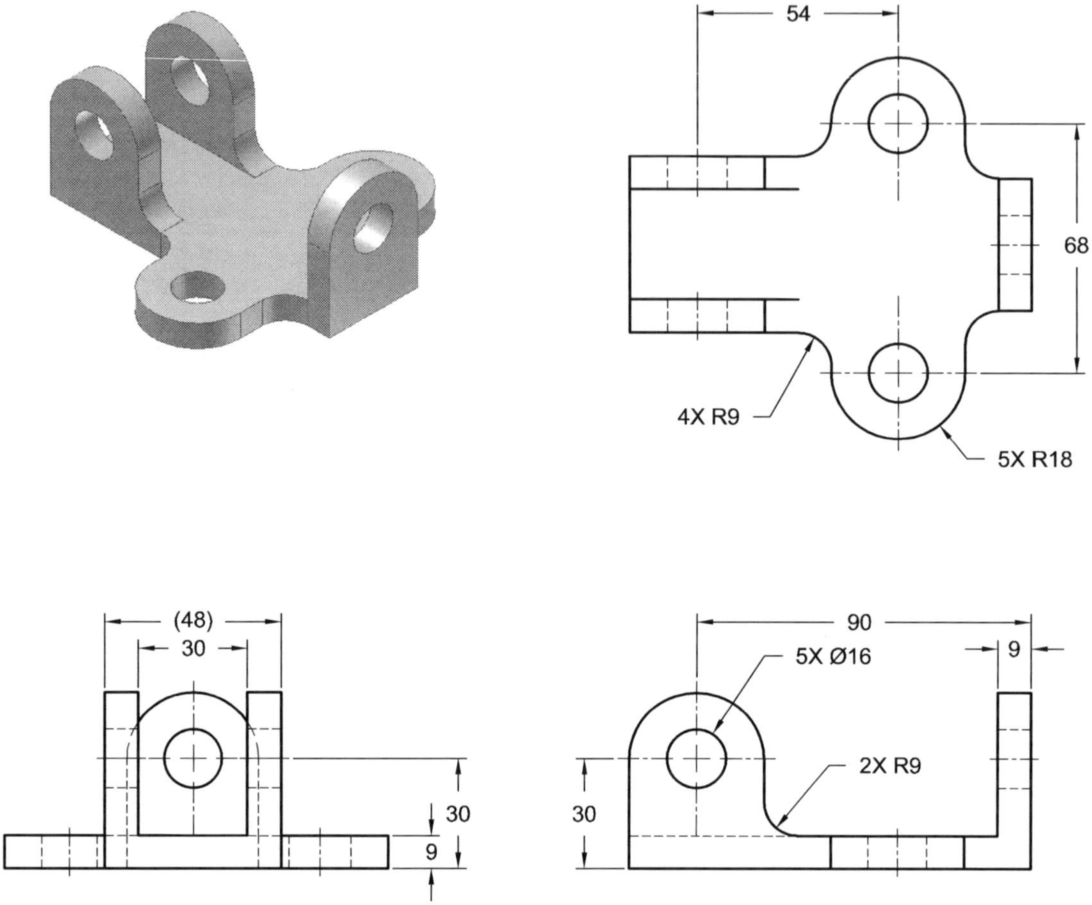

Figure 6-30 *Views and dimensions of the model*

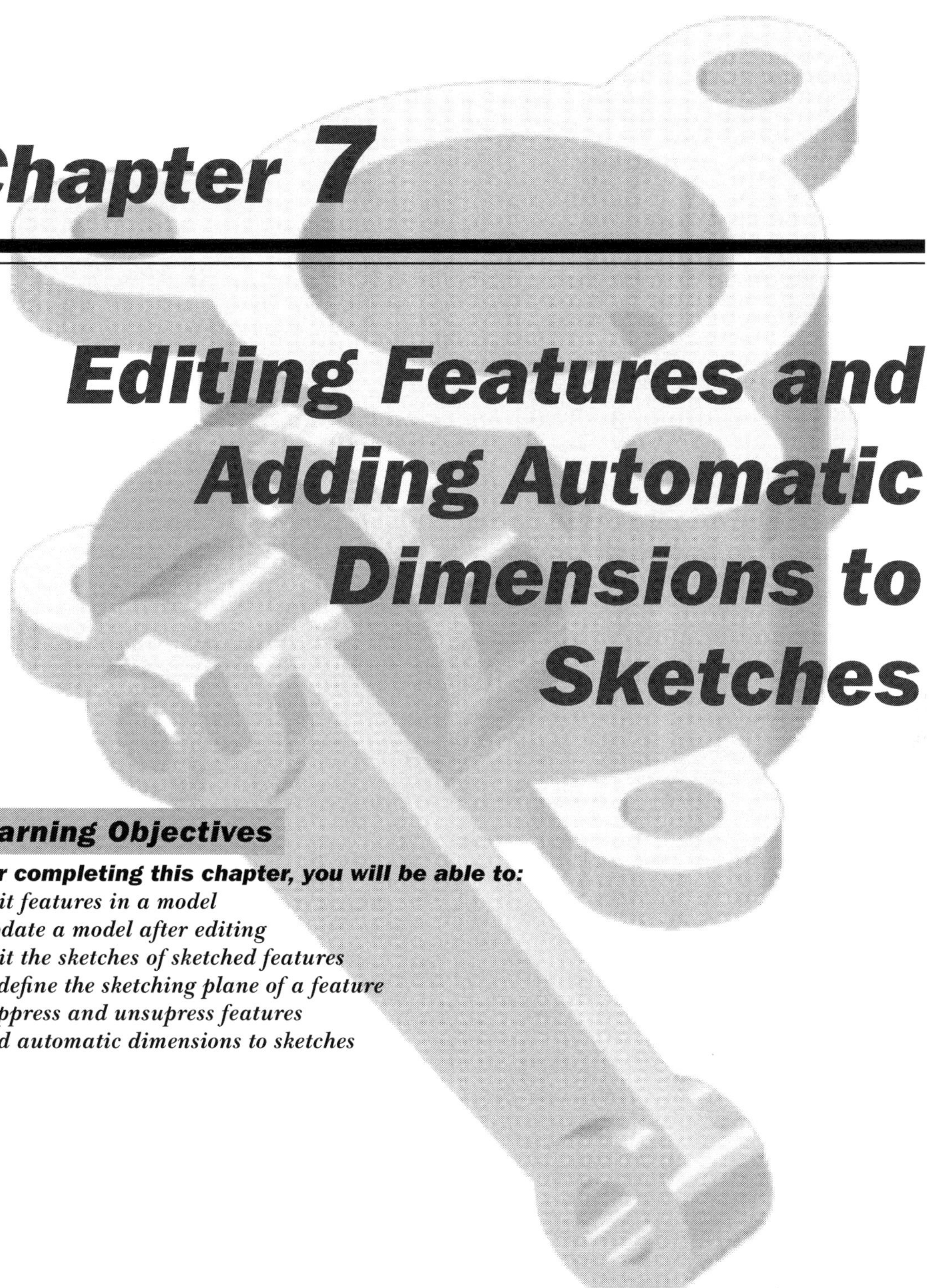

Chapter 7

Editing Features and Adding Automatic Dimensions to Sketches

Learning Objectives

After completing this chapter, you will be able to:

- *Edit features in a model*
- *Update a model after editing*
- *Edit the sketches of sketched features*
- *Redefine the sketching plane of a feature*
- *Suppress and unsupress features*
- *Add automatic dimensions to sketches*

CONCEPT OF EDITING FEATURES

Editing is one of the most important parts of designing. Most of the designs require editing either during or after their creation. As mentioned earlier, Autodesk Inventor is a feature-based solid modeling tool. As a result, the model created in Autodesk Inventor is a combination of various features. All these features are individual components and can be edited separately. This property gives this solid modeling software an edge over the other non-feature-based solid modeling tools. For example, Figure 7-1 shows a cylindrical part with six countersink holes created at some pitch circle diameter (PCD).

Now, in case you have to edit the features such that the countersink holes are to be changed into counterbore holes and the number of holes is to be increased, you just need to perform two editing operations. The first editing operation will open the **Holes** dialog box in which you can modify the countersink holes to counterbore holes. For this, you can specify various parameters for the counterbore hole in this dialog box. When you exit this dialog box, all the six countersink holes will be modified into counterbore holes. The second editing operation will open the **Circular Pattern** dialog box. In this dialog box, you can change the number of instances to eight, refer to Figure 7-2.

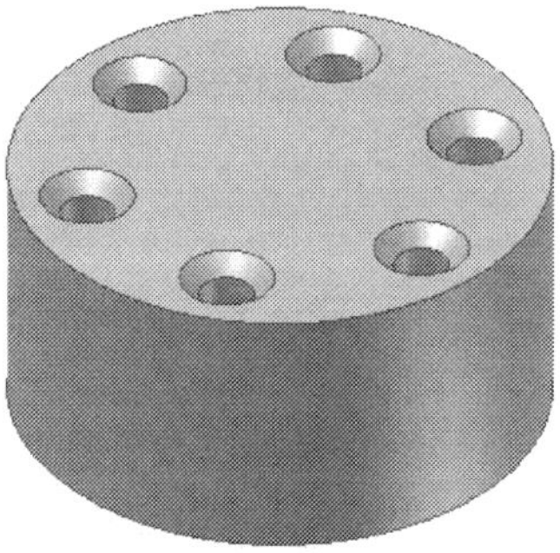

Figure 7-1 *Part with six countersink holes*

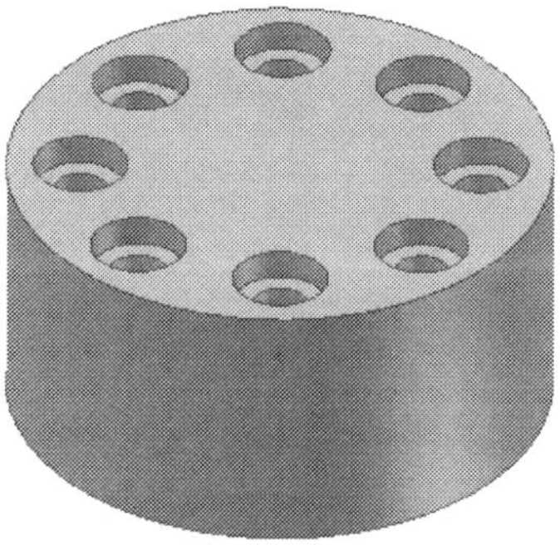

Figure 7-2 *Modified part with counterbore holes*

Similarly, you can also edit work features or sketches of the sketched features. The features created using the work features will be modified automatically when you edit the work features. For example, if you have created a feature on a work plane that is at an offset of 100 mm, the feature will be automatically repositioned on the change in the offset value of the work plane. In Autodesk Inventor, all the editing operations are performed using the **Browser Bar**.

Updating Edited Features

If you edit a feature using the **Browser Bar**, you do not have to update the feature to view the effect of the editing operation. This is because as soon as you exit the editing operation, the feature is automatically updated. However, if you modify the feature using dimensions, you will have to update the feature manually. Until the feature is updated after editing, it will not display the modified values. To update the feature with the modified values, choose the **Local Update** button in the **Quick Access Toolbar**. This button will be activated when you modify the dimensions of any features.

Redefining the Sketching Plane of a Sketched Feature

Sometimes, you may need to relocate a feature drawn on one of the planes to another plane. For example, you may need to relocate a cylinder drawn on the XZ plane to the YZ plane. Autodesk Inventor allows you to relocate features on other planes by redefining the sketching plane. After redefining it, the necessary changes are automatically made in the orientation of the model. For example, a cylinder drawn on the XZ plane stands vertically. However, the same cylinder drawn on the YZ plane lies horizontally. You can select work planes in the graphics window. Alternatively, you can expand the **Origin** tab in the **Browser Bar** and then choose the desired plane in which you want the model to be reoriented.

To redefine the sketching plane of a sketched feature, click on the + sign located on the left of the sketched feature in the **Browser Bar**; the name of the sketch of the corresponding feature will appear below it in the **Browser Bar**. Right-click on the sketch and choose **Redefine** from the shortcut menu; you will be prompted to select a work plane or planar face to redefine the sketch. Select the new work plane or planar face for the sketched feature; the sketch of the feature will be relocated on the new plane and the model will reorient, based on the new parameters. Also, all the features created with reference to the current features will be updated automatically.

SUPPRESSING AND UNSUPPRESSING THE FEATURES

Sometimes, there may be a situation where you want that some of the features should not show up in the drawing views of the model or in the printout of the model. In any of the non-feature based solid modeling tools, you will have to either delete the feature or create it after taking the printout. However, in Autodesk Inventor, you can simply suppress the feature not required. Once the feature is suppressed, it will neither be displayed in the drawing views nor in the printout of the model. Remember that in such cases the features are not deleted, they are temporarily turned off. Note that all the features that are dependent on the feature that you select are also suppressed. To suppress a feature, right-click on it in the **Browser Bar** and then choose **Suppress Features** from the shortcut menu.

ADDING AUTOMATIC DIMENSIONS TO SKETCHES

Ribbon: Sketch > Constrain > Automatic Dimensions and Constraints

Autodesk Inventor allows you to add dimensions and constraints automatically. Note that if you cannot apply all dimensions and constraints required in a sketch. The dimensions are used in association with the general dimensions to fully constrain the sketch. In Autodesk Inventor, automatic dimensions are added using the **Automatic Dimensions and Constraints** tool.

PROJECTING ENTITIES IN THE SKETCHING ENVIRONMENT

Autodesk Inventor allows you to project the edges of an existing feature to a sketching plane while drawing the sketches. The projected edges are converted into sketched entities and can be used as a part of the sketch. You can project the selected edge or face of a feature, or project the part of the model that is cut by the sketching plane. You can project the entities using various tools available in the **Project Geometry** drop-down which is available in the **Create** panel in the **Sketch** tab.

Projecting Edges or Faces

Ribbon:	Sketch > Create > Project Geometry drop-down > Project Geometry

You can project the selected edges or faces of a feature on a sketching plane by choosing the **Project Geometry** tool from the **Create** panel of the **Sketch** tab.

Projecting Cutting Edges

Ribbon:	Sketch > Create > Project Geometry drop-down > Project Cut Edges

The cutting edges are meant to define the contour of the model that is created when you define a sketching plane on the face of a model or inside the model. When you define a sketching plane inside the model, it cuts the model, thus forming cutting edges. You can project these cutting edges by choosing the **Project Cut Edges** tool from the **Create** panel of the **Sketch** tab.

Projecting 2D Sketch on a 3D Face

Ribbon:	Sketch > Create > Project Geometry drop-down > Project to 3D Sketch

With the introduction of the **Project to 3D Sketch** tool in Autodesk Inventor, you can now project a 2D sketch onto a 3D face. To do so, invoke the **Project to 3D Sketch** tool; the **Project to 3D Sketch** dialog box will be displayed. In this dialog box, the **Faces** button is activated by default. If not, then select the **Project** check box; the **Faces** button will be activated. If you move the cursor over the faces of the model, they will be highlighted. Select the faces on which you want to project the 2D sketch; the sketch will be projected on the selected faces.

Projecting DWG Geometry

Ribbon:	Sketch > Create > Project Geometry drop-down > Project DWG Geometry

You can project a DWG geometry by using the **Project DWG Geometry** tool from the **Create** panel of the **Sketch** tab. On invoking this tool, you will be prompted to select a DWG line, arc, polyline or other single geometry. Select the geometry that you want to project by choosing the corresponding option from the mini toolbar; the selected geometry will be projected on the sketching plane or face.

TUTORIALS

Tutorial 1

In this tutorial, you will create the model of the Gear-Shifter Link shown in Figure 7-3. Its views and dimensions are shown in the same figure. **(Expected time: 45 min)**

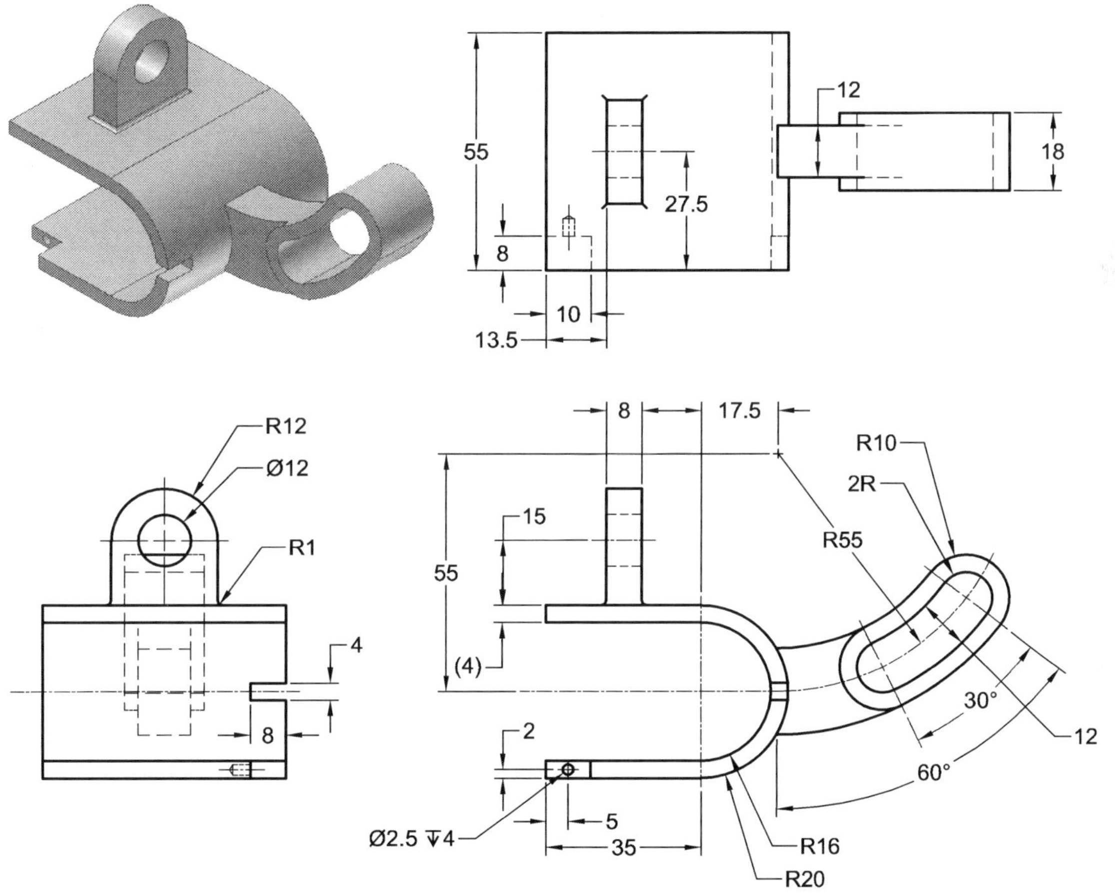

Figure 7-3 *Views and dimensions for Tutorial 1*

The following steps are required to complete this tutorial:

a. Create the base feature which is a reverse C-like feature. Its sketch will be created on the XY plane and extruded using the **Symmetric** option.
b. Define a new sketch plane on the XY plane and add the first join feature to the circular face of the base feature.
c. Again, define a new sketch plane on the XY plane and draw the sketch for the second join feature. Extrude this feature using the **Symmetric** option.
d. Define a new sketch plane on the front face of the second join feature and draw the sketch for the cut feature. Extrude this sketch using the **Cut** operation.

e. Suppress all features, except the base feature, and then define a work plane at an offset of 17.5 mm from the left face of the base feature. Draw the sketch for the third join feature and extrude it using the **Symmetric** option.
f. Add fillet to the third join feature.
g. Suppress the last two features and create the slots and hole on the front face of the base feature.
h. Finally, unsuppress all features to complete the model.

Creating the Base Feature

1. Start a new metric standard part file and then draw the sketch of the base feature on the XY plane, as shown in Figure 7-4.

2. Exit the sketching environment and then extrude the sketch to a distance of 55 mm using the **Symmetric** option. The base feature of the model is shown in Figure 7-5.

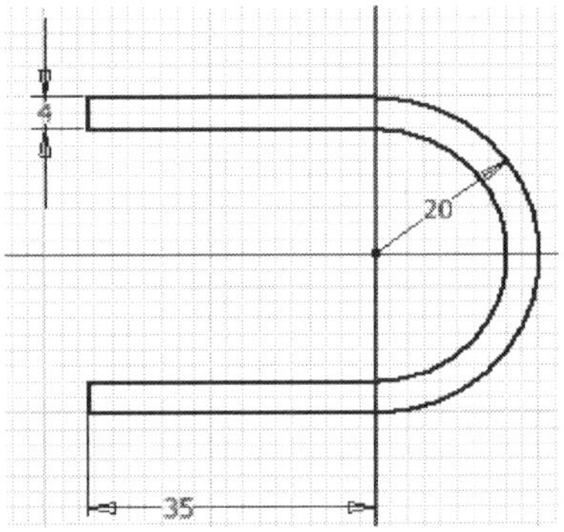

Figure 7-4 *Sketch of the base feature*

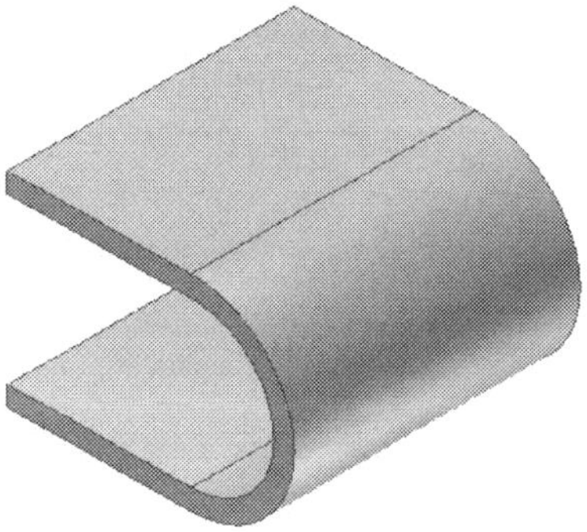

Figure 7-5 *Base feature of the model*

Creating Curved Features

As the base feature was created on the XY plane and was extruded using the **Symmetric** option, you can create the sketches for the curved features on the same plane and then extrude them as required using the **Symmetric** option.

1. Draw the sketch of the curved feature on the XY plane. Add the required constraints and dimensions to it. The sketch of the first join feature after applying all dimensions and constraints is shown in Figure 7-6.

2. Exit the sketching environment and then extrude the sketch of the first join feature to a distance of 12 mm using the **Join** operation. Use the **Symmetric** option for creating the feature. The isometric view of the model is shown in Figure 7-7.

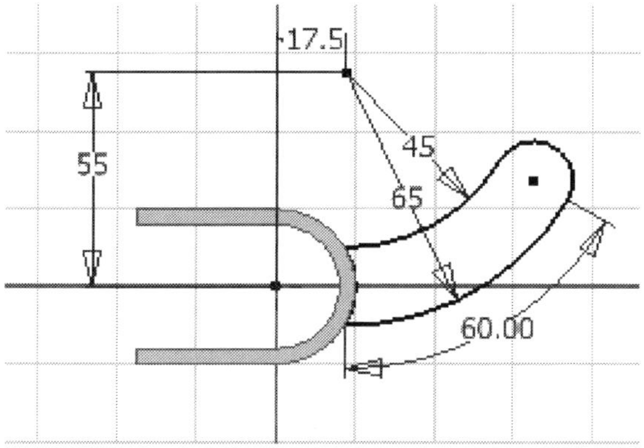

Figure 7-6 *Sketch of the first join feature*

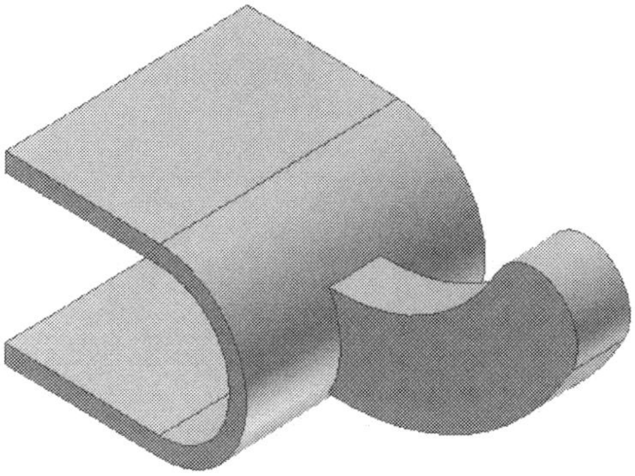

Figure 7-7 *The isometric view of the model*

3. Draw the sketch of the second join feature on the XY plane, as shown in Figure 7-8. After drawing the sketch, apply the Concentric and Equal constraints to it and to the existing feature.

> **Note**
> *You can use the **Slice Graphics** option for slicing a feature to create the sketch of the second join feature. This option can be invoked from the Marking menu that is displayed when you right-click in the graphics window in the sketching environment. Alternatively, press the F7 key to slice graphics.*

4. Exit the sketching environment and then extrude the sketch to a distance of 18 mm using the **Join** operation and the **Symmetric** option, see Figure 7-9.

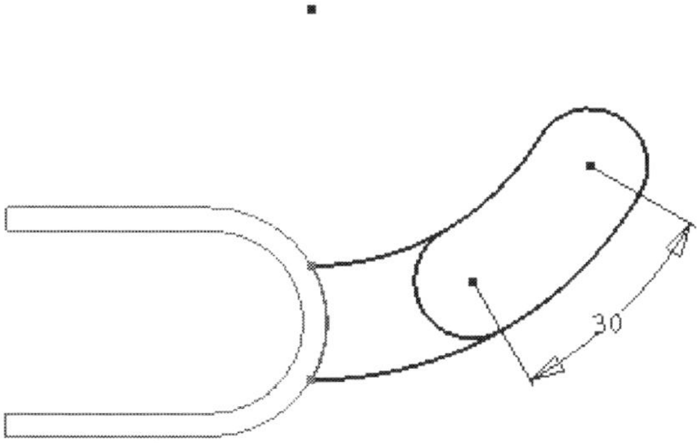

Figure 7-8 *Sketch of the second join feature in the wireframe display*

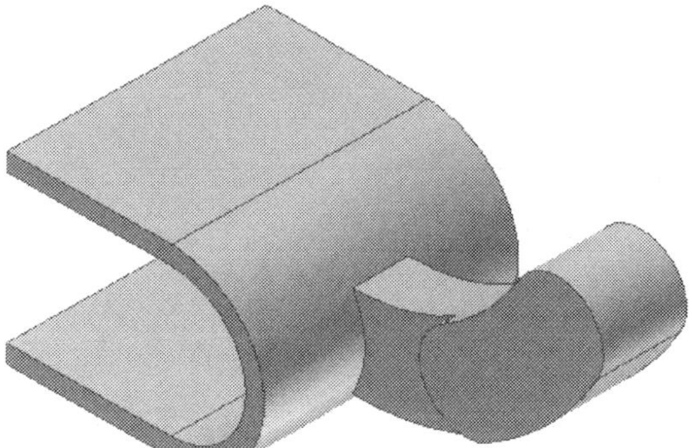

Figure 7-9 *Model after creating the second join feature*

5. Specify a new sketch plane on the front face of the second join feature and then create the sketch of the cut feature on this plane. If required, you can project the existing sketch. You can offset and dimension the reference entities to create the sketch of the cut feature, as shown in Figure 7-10.

6. Exit the sketching environment and then click on the sketch; a mini toolbar is displayed.

7. Choose the **Create Extrude** button from the mini toolbar; the **Properties-Extrude** dialog box is displayed.

8. Click inside the sketch created for the cut feature and then choose the **Cut** button from the **Boolean** area of the **Output** node.

9. Choose the **Through All** button from the **Behavior** node and choose the **OK** button in the dialog box; the cut feature is created, as shown in Figure 7-11. Note that if you are not getting required result then you need to choose the **Flipped** button in the **Direction** area.

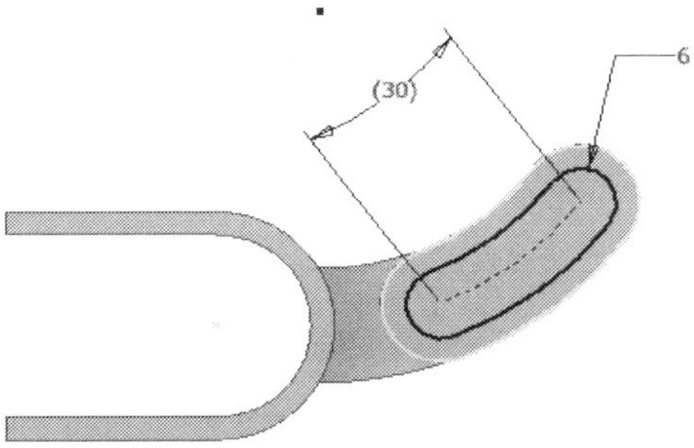

Figure 7-10 *Sketch of the cut feature*

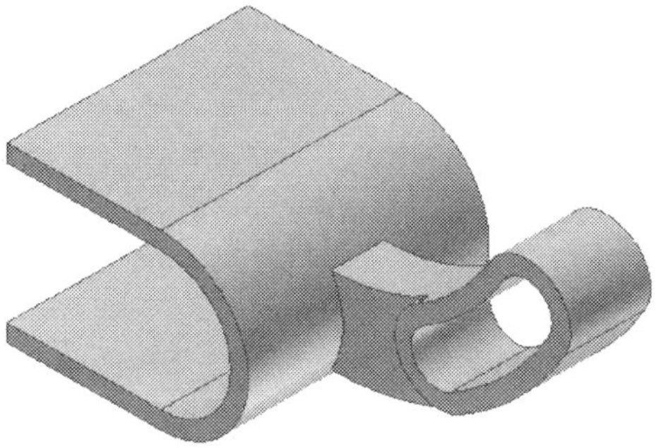

Figure 7-11 *Model after creating the cut feature*

Suppressing Features

As mentioned earlier, in a complex model, it is better to suppress the features not required while creating a particular feature. After creating all the features, you can unsuppress the suppressed features. In Autodesk Inventor, the features are suppressed using the **Browser Bar**.

1. Right-click on **Extrusion2** (first join feature) in the **Browser Bar** to display a shortcut menu.

2. Choose **Suppress Features** from the shortcut menu displayed; the **Autodesk Inventor Professional-Suppress Feature** dialog box is displayed. Also, the first join feature is no more visible. However, the second join feature and the cut feature remain visible on the model.

3. Choose **Accept** from the **Autodesk Inventor Professional - Suppress Feature** dialog box. Now, right-click on **Extrusion3** (second join feature) in the **Browser Bar** and then choose **Suppress Features** from the shortcut menu displayed to suppress the second join feature and the cut feature.

   The only feature that is visible now is the base feature of the model.

Creating the Third Join Feature

The third join feature is created on an offset work plane. This work plane will be at an offset distance of -17.5 mm from the left face of the base feature. Negative value of offset distance will ensure that the work plane is offset inside the model.

1. Choose the **Offset from Plane** tool from **3D Model > Work Features > Plane** drop-down and define a new work plane at an offset of -17.5 mm from the left face of the base feature. Create the sketch for the third join feature on the new plane.

2. Create a circle inside the sketch so that when you extrude the sketch, a hole is also created, see Figure 7-12.

3. Extrude the sketch upto a distance of 8 mm using the **Join** operation and the **Symmetric** option. If the **Autodesk Inventor** warning window is displayed, choose **Accept** from it. The model after creating the third join feature is displayed in Figure 7-13.

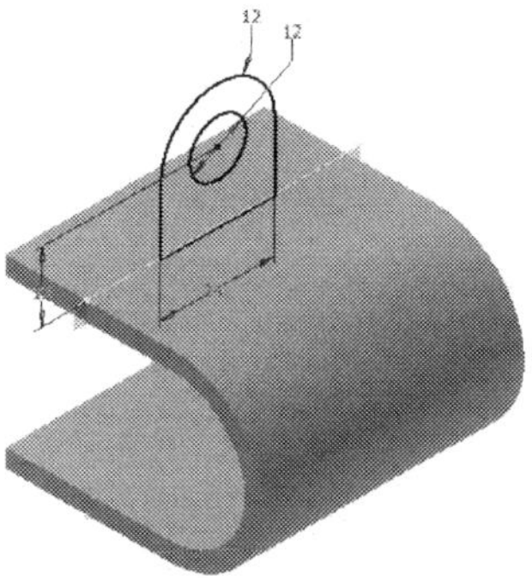

Figure 7-12 *Sketch for the third join feature*

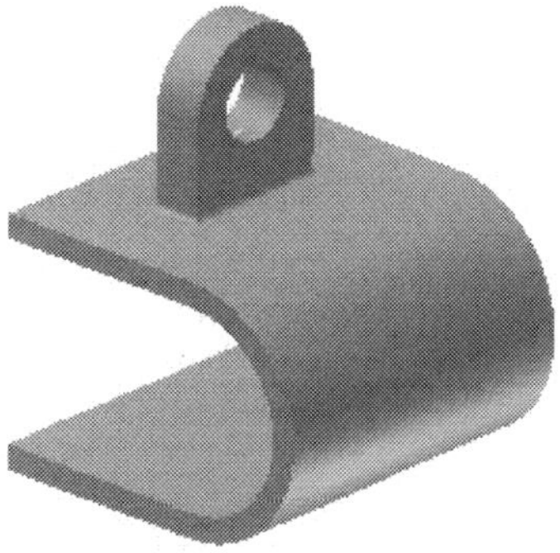

Figure 7-13 *Model after creating the third join feature*

Creating the Fillet Feature on the Third Join Feature

1. Invoke the **Fillet** tool from the Marking menu.

2. Select the **Loop** radio button from the **Select mode** area and enter **1** in the **Radius** edit box; you are prompted to select the loop on the model.

3. Select the loop on the third join feature of the model, as shown in Figure 7-14. Next, choose **OK** from the **Fillet** dialog box; the fillet feature is created, as shown in Figure 7-15.

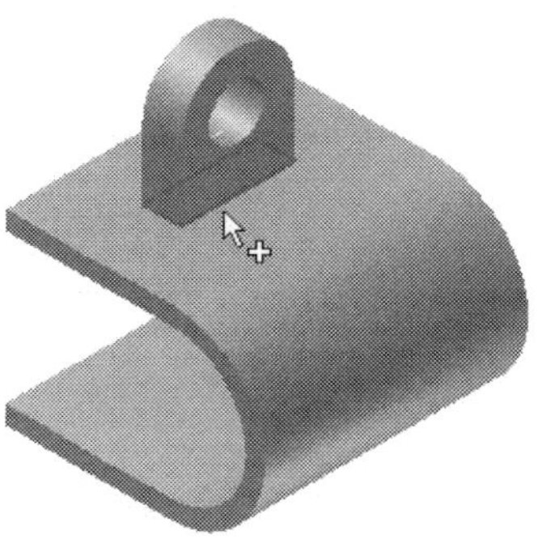

Figure 7-14 *Loop selected for creating the fillet*

Figure 7-15 *Model after creating the fillet on the third join feature*

Suppressing the Third Join Feature

1. Right-click on **Extrusion5** (third join feature) in the **Browser Bar** to display a shortcut menu. Choose **Suppress Features** from the shortcut menu; the fillet feature is suppressed because it is dependent on the third join feature. Now, the only visible feature is the base feature.

Creating Slots and the Hole

In this section, the slots will be created on the front face of the base feature. As both the slots will be cut to the same distance, you can draw the sketch for both the slots together and then extrude them using the **Cut** operation.

1. Define a new sketch plane on the front face of the base feature.
2. Draw the sketch for both slots, as shown in Figure 7-16.
3. Invoke the **Extrude** tool and then create the slots of depth 8 mm using the **Cut** operation, as shown in Figure 7-17.

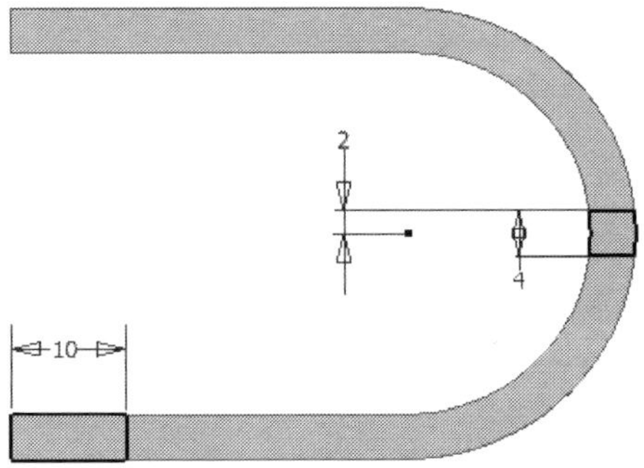

Figure 7-16 *Sketch for slots*

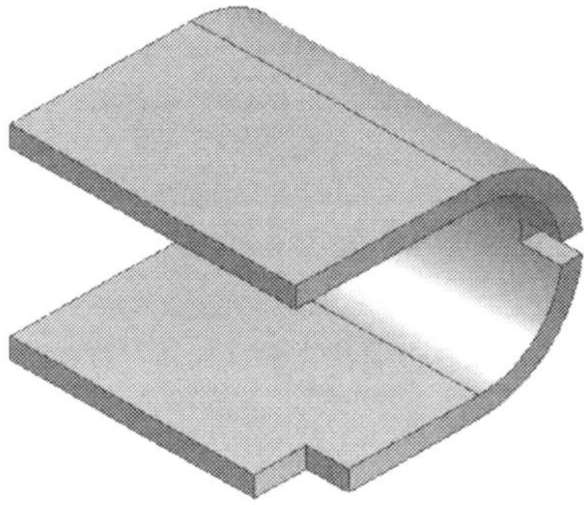

Figure 7-17 *Model after creating slots*

4. Create a drilled hole of 2.5 mm diameter on the left face of the slot, refer to Figure 7-18. For the location of the hole, refer to Figure 7-3. The depth of the hole is 4 mm. The model after creating the hole is shown in Figure 7-18.

Unsuppressing Features

Once all the features of the model have been created, you can unsuppress the suppressed features and save the model. As you know, all the suppressed features will be displayed in light gray color and will have a line striking out their names in the **Browser Bar**.

1. Right-click on **Extrusion2** (first join feature) in the **Browser Bar** and then choose **Unsuppress Features** from the shortcut menu displayed.

   If the Autodesk Inventor encounters an error while updating the features, the **Autodesk Inventor Professional - Unsuppress Feature** dialog box is displayed. Choose **Accept** from the dialog box displayed; the second join feature is unsuppressed.

2. Similarly, unsuppress the remaining suppressed features. Choose **Accept** in the **Autodesk Inventor Professional -Unsuppress Feature** dialog box, whenever it is displayed. The final solid model of the Gear-shifter link after unsuppressing all the suppressed features is shown in Figure 7-19.

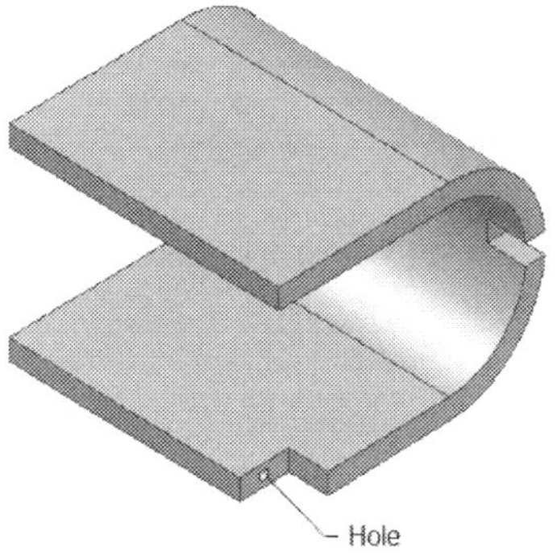

Figure 7-18 *Model after creating the hole*

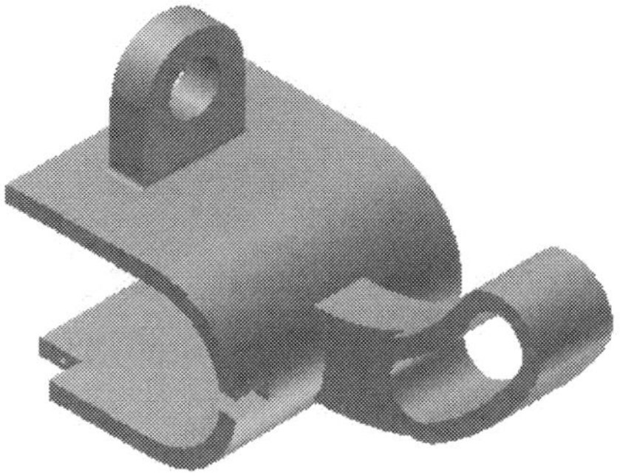

Figure 7-19 *Final solid model of the Gear-shifter link*

Saving the Model

1. Save the model with the name *Tutorial1* at the location *C:\Inventor_2020|c07* and then close the file.

Tutorial 2

In this tutorial, you will create the model shown in Figure 7-20. Its views and dimensions are shown in the same figure. **(Expected time: 45 min)**

The following steps are required to complete this tutorial:

a. Create the base feature on the XY plane. The sketch for the base feature consists of a square with fillets on all four corners.
b. On the front face of the base feature, create counterbore holes by using the center points of fillets as the center of holes.
c. Suppress the holes and create the cylindrical join feature on the front face of the base feature.
d. Add two rectangular join features to the cylindrical feature and create the rectangular cut feature on one of the rectangular join features.
e. Create drilled holes by defining the sketch plane on the required planes. Once all features are created, unsuppress the holes on the base feature.
f. Finally, create the fillet of radius 5 mm.

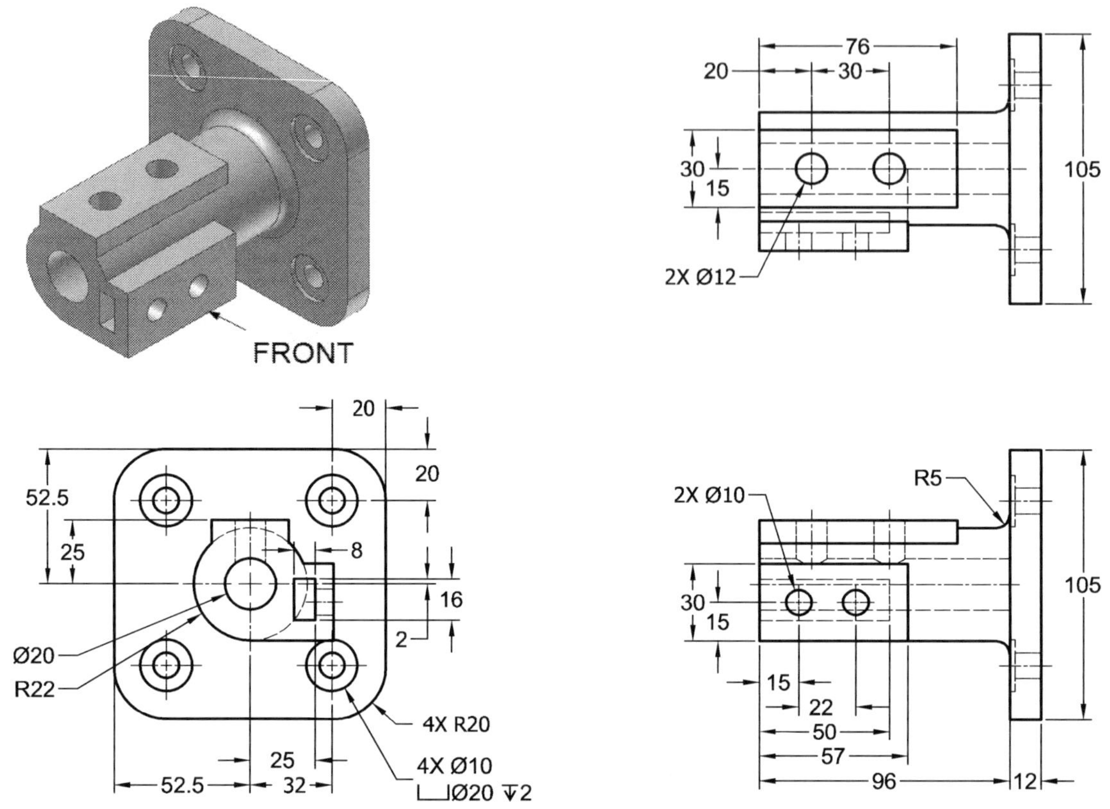

Figure 7-20 *Views and dimensions for Tutorial 2*

Creating the Base Feature

1. Start a new metric part file and then draw the sketch of the base feature on the XY plane.

   The sketch of the base feature will be a square of side 105 mm with all the four corners filleted with a radius of 20 mm.

2. Exit the sketching environment and extrude the sketch upto a distance of 12 mm. The base feature of the model is shown in Figure 7-21.

Creating the Holes

The base feature has four counterbore holes. You can create these holes concentric with the cylindrical faces of the fillets at the corners. Alternatively, you can create one of the holes and then create a rectangular pattern for creating the remaining three holes.

1. Invoke the **Hole** tool and then create four counterbore holes using on the cylindrical faces. Refer to Figure 7-20, for dimensions.

The model after creating the counterbore holes is shown in Figure 7-22.

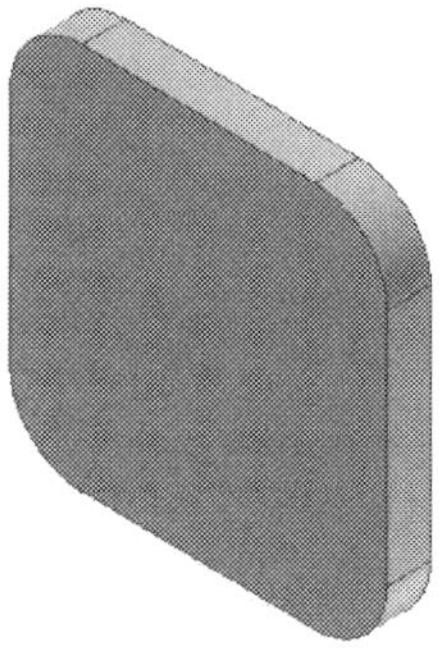

Figure 7-21 Base feature of the model

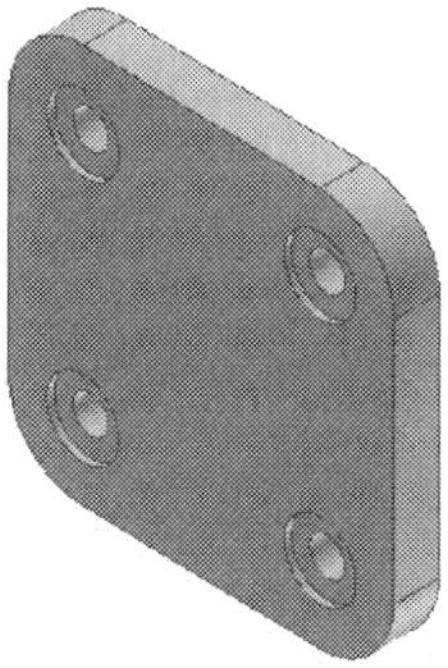

Figure 7-22 Model after creating counterbore holes

Note

*You need to use the **Through All** option from the **Termination** area to create through holes.*

Suppressing Holes

As you have created four holes, the **Browser Bar** will display these holes with names **Hole1**, **Hole2**, **Hole3**, and **Hole4**. You need to select all the four holes and suppress them.

1. Press and hold the SHIFT or CTRL key and select all four holes from the **Browser Bar**.

2. Right-click on the selected holes in the **Browser Bar** to display a shortcut menu. From the shortcut menu, choose **Suppress Features**; all the four holes get suppressed. The only feature that is visible is the base feature.

Creating Cylindrical and Rectangular Join Features

1. Define a new sketch plane on the front face of the base feature and then draw a circle of 44 mm diameter, as shown in Figure 7-23. Exit the sketching environment.

2. Invoke the **Extrude** tool and then extrude the circle upto a distance of 96 mm, as shown in Figure 7-24.

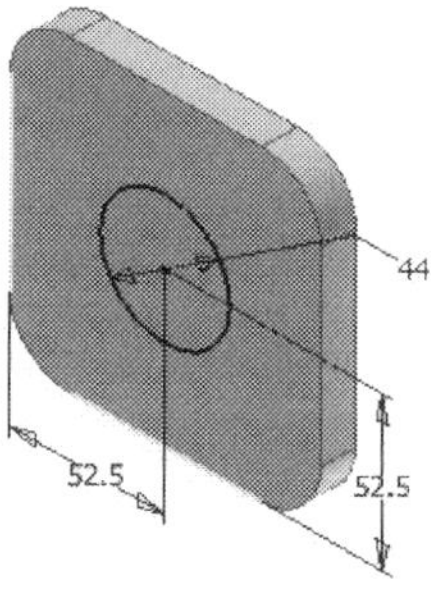

Figure 7-23 Sketch for the cylindrical feature

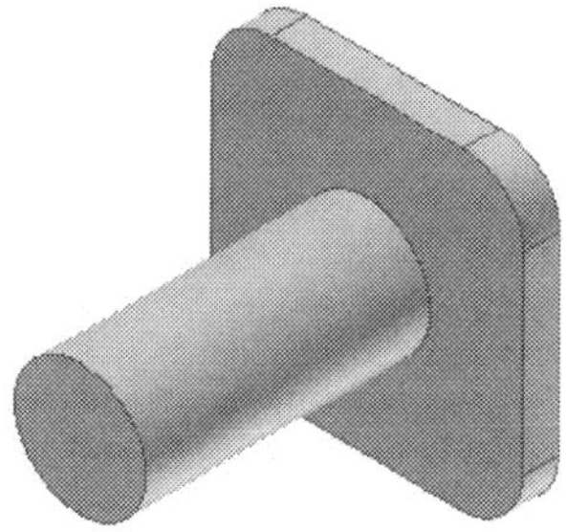

Figure 7-24 Cylindrical feature created

3. Define a new sketch plane on the front face of the cylindrical feature and create the sketch of the first rectangular feature, as shown in Figure 7-25. Exit the sketching environment.

4. Invoke the **Extrude** tool and then extrude the sketch upto a distance of 76 mm, as shown in Figure 7-26.

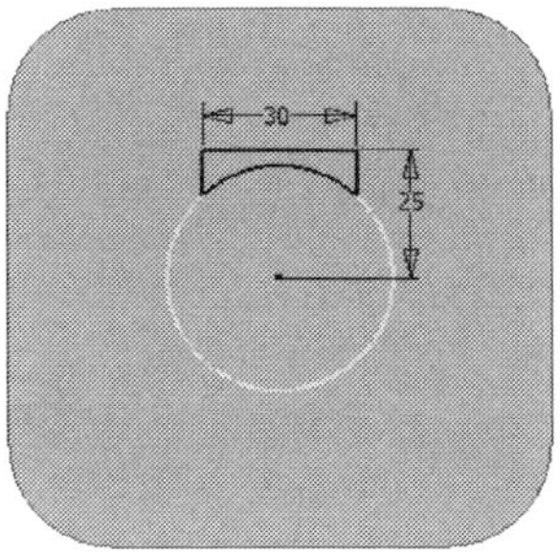

Figure 7-25 *Sketch of the first rectangular feature*

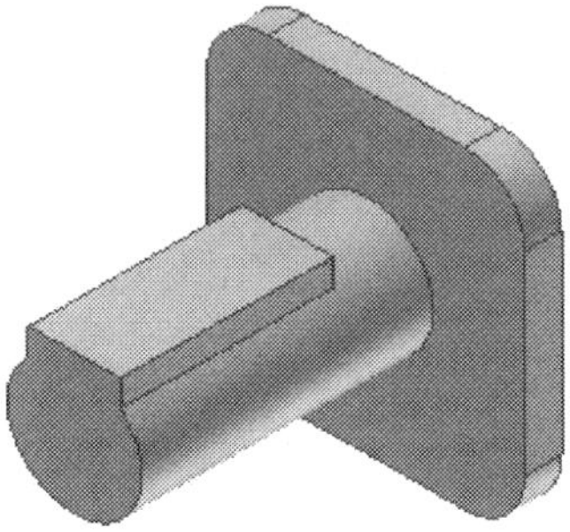

Figure 7-26 *Model after creating the first rectangular feature*

5. Similarly, create the sketch of the second rectangular feature and extrude it upto a distance of 57 mm, refer to Figures 7-27 and 7-28.

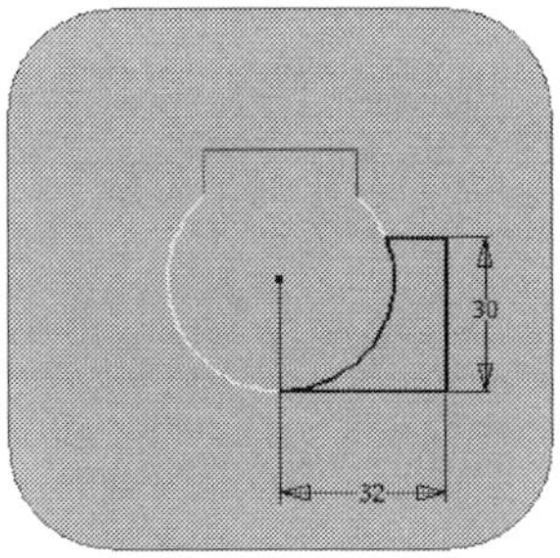

Figure 7-27 *Sketch of the second rectangular feature*

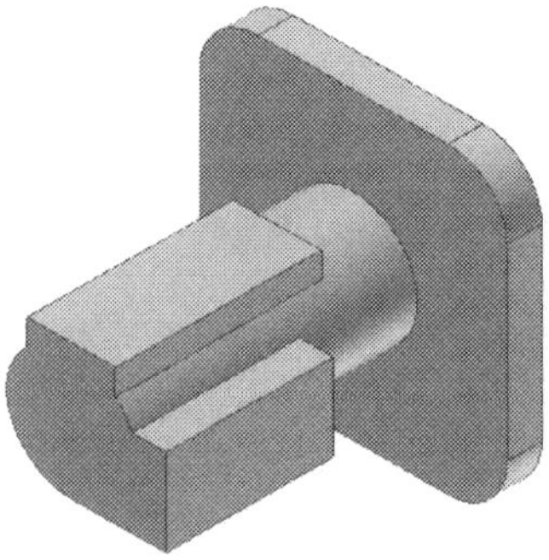

Figure 7-28 *Model after creating the second rectangular feature*

Creating the Cut Feature and Holes

1. Define a new sketch plane on the front face of the cylindrical feature and then create a rectangular cut feature of depth 50 mm, refer to Figures 7-29 and 7-30.

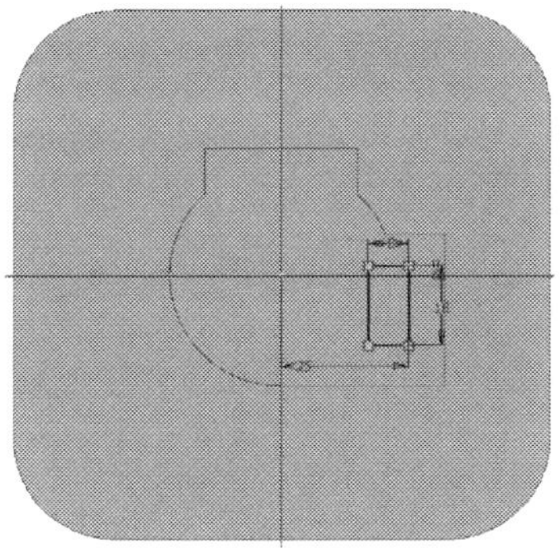

Figure 7-29 Sketch of the rectangular cut feature

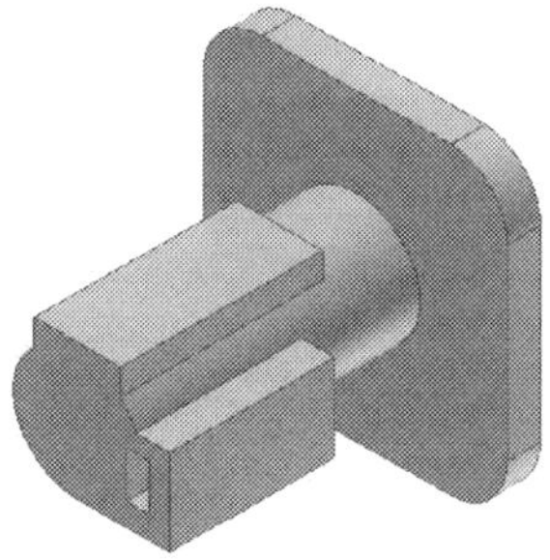

Figure 7-30 Model after creating the rectangular cut feature

2. Create a hole of 20 mm diameter and 96 mm depth on the front face of the cylindrical feature. The model after creating the hole and the cut feature is shown in Figure 7-31.

3. Similarly, create holes one by one on the faces of the rectangular join features, see Figure 7-32. For the dimensions and location of holes, refer to Figure 7-20.

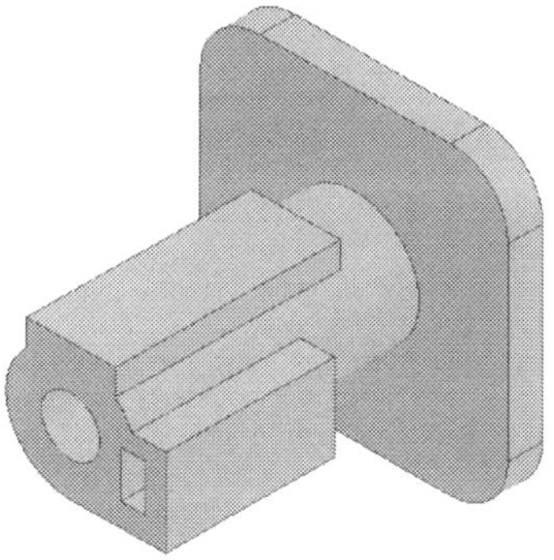

Figure 7-31 Model after creating holes and the cut feature

Unsuppressing Counterbore Holes

1. From the **Browser Bar**, select all the holes created on the base feature. Right-click on the selected holes and choose **Unsuppress Features** from the shortcut menu displayed; all the counterbore holes will again be displayed on the base feature.

Creating the Fillet and Saving the File

1. Invoke the **Fillet** tool and create a fillet of radius 5 mm on the circular edge created between the base feature and the cylindrical join feature. The final model for Tutorial 2 after unsuppressing the counterbore holes and creating the fillet is shown in Figure 7-32.

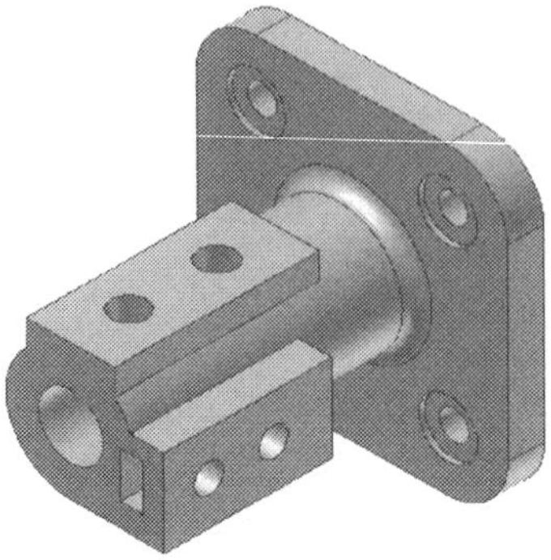

Figure 7-32 *Final model for Tutorial 2*

2. Save the model with the name *Tutorial2* at the location *C:\Inventor_2020\c07* and then close the file.

EXERCISES

Exercise 1

Create the model of the Slide Bracket shown in Figure 7-33. Its views and dimensions are shown in the same figure. **(Expected time: 45 min)**

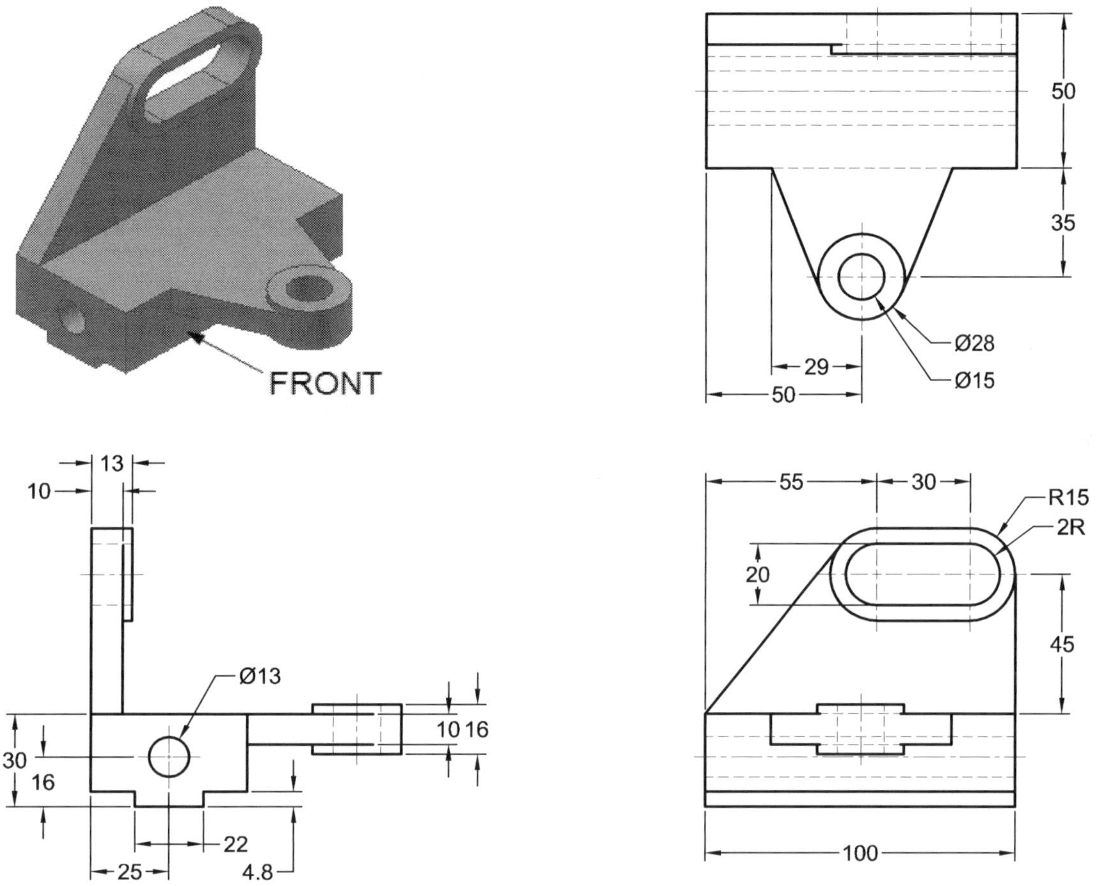

Figure 7-33 *The model, its views and dimensions*

Exercise 2

Create the model shown in Figure 7-34. Its views and dimensions are shown in the same figure. **(Expected time: 45 min)**

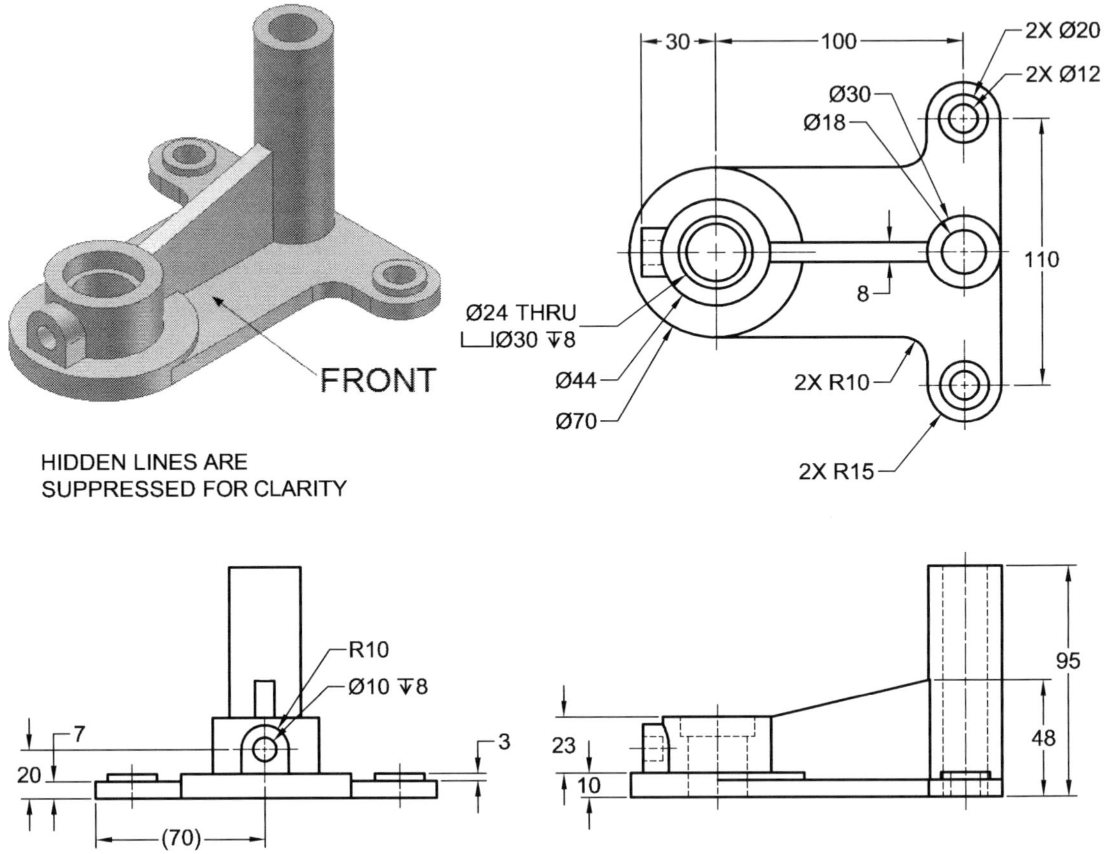

Figure 7-34 *The model, its views and dimensions*

Chapter 8

Advanced Modeling Tools-II

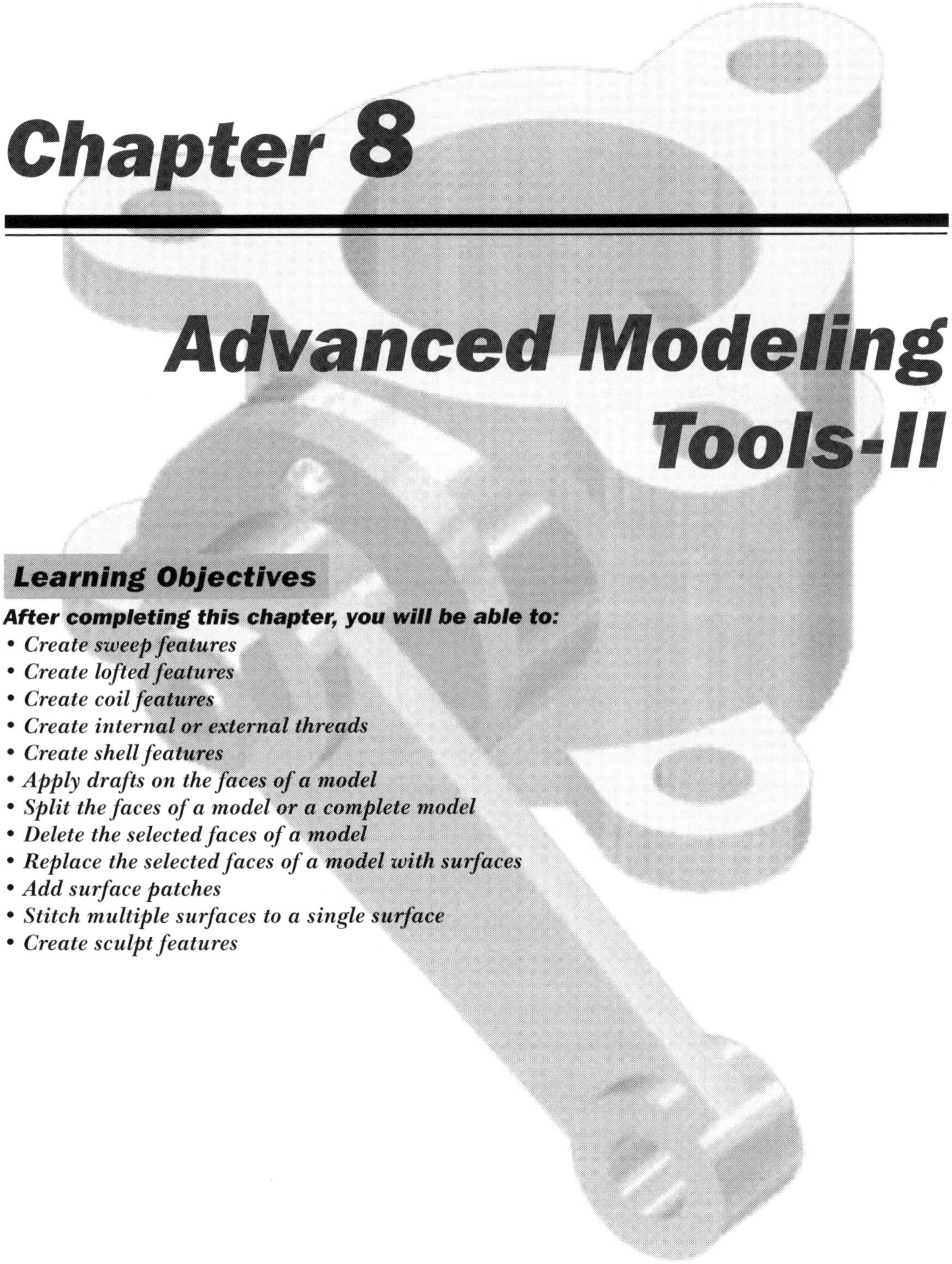

Learning Objectives

After completing this chapter, you will be able to:

- *Create sweep features*
- *Create lofted features*
- *Create coil features*
- *Create internal or external threads*
- *Create shell features*
- *Apply drafts on the faces of a model*
- *Split the faces of a model or a complete model*
- *Delete the selected faces of a model*
- *Replace the selected faces of a model with surfaces*
- *Add surface patches*
- *Stitch multiple surfaces to a single surface*
- *Create sculpt features*

ADVANCED MODELING TOOLS

The first few advanced modeling tools are discussed in Chapter 6, Advanced Modeling Tools-I. In this chapter, you will learn about the remaining tools.

Creating Sweep Features

Ribbon: 3D Model > Create > Sweep

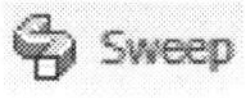

You can create a sweep feature by using the **Sweep** tool. A sweep feature is created when a closed sketch is swept along an open or a closed path. Therefore, to create a sweep feature, two unconsumed sketches are required which are a closed sketch (called profile) and a path.

Creating Lofted Features

Ribbon: 3D Model > Create > Loft

Lofted features are created by blending more than one geometry together. The geometries may or may not be parallel to each other. The sketches for the solid loft features should be closed profiles or a point. However, for a surface model, the sketches can be open profiles.

Creating Coil Features

Ribbon: 3D Model > Create > Coil

A coil feature is created by sweeping a profile about a helical path. The examples of coil feature are springs, filaments of light bulbs, and so on. To create a coil feature, you need a profile and an axis. You can also select the standard X, Y, Z axes, or a new work axis to create the coil feature. Coil features are created using various tabs from the **Coil** dialog box. This dialog box is invoked by choosing the **Coil** tool from the **Create** panel of the **3D Model** tab. Depending on the parameters specified in the **Coil** dialog box, an imaginary helical path is created and the profile is swept along that path. Therefore, to create a coil feature, you need only one unconsumed sketch, which defines the profile of the coil section.

Creating Threads

Ribbon: 3D Model > Modify > Thread

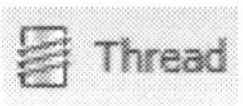

Autodesk Inventor allows you to create internal or external threads directly on a model. Internal threads are created on the inner surface of a feature. You can create threads using the **Thread** tool.

Creating Shell Features

Ribbon: 3D Model > Modify > Shell

Shelling is a process of scooping out material from a model to make it hollow. The resulting model will be a structure of walls with cavity. You can also remove some of the faces of the model or apply different wall thicknesses to some of the faces.

Applying Drafts

Ribbon: 3D Model > Modify > Draft

Face draft is a process of tapering the outer faces of a model for its easy removal from casting during manufacturing. You can add a face draft using the **Draft** tool.

Creating Split Features

Ribbon: 3D Model > Modify > Split

In Autodesk Inventor, the **Split** tool can be used for splitting the entire part or the faces of the part.

Deleting Faces

Ribbon: 3D Model > Modify > Delete Face

Autodesk Inventor allows you to delete one or more planar faces or non-planar lumps in a model or in a surface. Depending on the face selected to be deleted, the resulting model is converted into a surface. You can also force the adjacent faces to extend and intersect such that they heal the surface.

Replacing Faces with Surfaces

Ribbon: 3D Model > Surface > Replace Face

Autodesk Inventor allows you to replace the selected faces of a model with one or more selected surfaces or work planes. Note that the surface must intersect the complete face that you want to replace.

Creating Planar Boundary Patches

Ribbon: 3D Model > Surface > Boundary Patch

Autodesk Inventor allows you to create planar boundary patches on one or more closed loops or edges using the **Boundary Patch** tool.

Stitching Surfaces

Ribbon: 3D Model > Surface > Stitch Surface

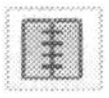

Sometimes, while splitting parts, you may need to use more than one surface as the splitting tool. The **Split** tool allows you to select only one surface to split parts or faces. In such cases, you can join more than one surface together so that they form a single surface.

Working with the Sculpt Tool

Ribbon: 3D Model > Surface > Sculpt

The **Sculpt** tool is used to add or remove material from an existing model by using a surface or a datum plane.

Working with the Bend Part Tool

Ribbon: 3D Model > Modify > Bend Part

The **Bend Part** tool is used to bend components or portions of components by using different options. To bend a component, first you need to sketch a line about which the component will be bent. This line is called the Bend Line. It can also be defined as the tangency line at which the component transforms into a bend. After specifying the tangency conditions between the bend line and the component, you can define the side of the component to be bent, the direction and angle of the bend, and other parameters.

TUTORIALS

Tutorial 1

In this tutorial, you will create the model shown in Figure 8-1. Its dimensions are given in Figure 8-2. **(Expected time: 45 min)**

The following steps are required to complete this tutorial:

a. The base feature of the model is a sweep feature. Create the path of the sweep feature on the XY plane. Next, define a work plane normal to the path and position it at the start point of the path. Create the profile of the sweep feature on this work plane. Use the **Sweep** tool to create the sweep feature.
b. Create the inner cavity using the **Shell** tool.
c. Add the remaining features (join features, drilled holes and their patterns, and counterbore hole) on both ends of the sweep feature.

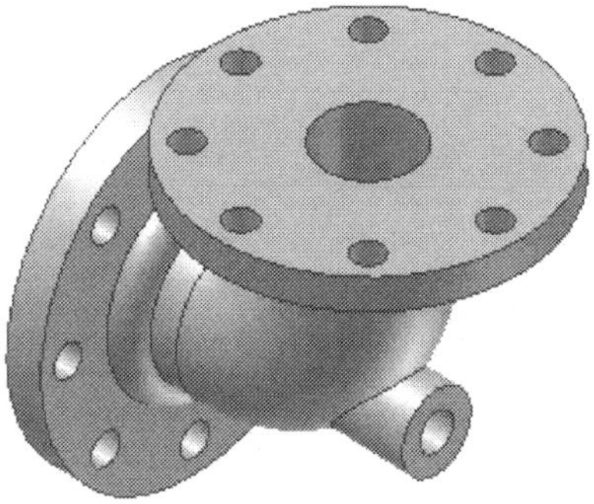

Figure 8-1 *Model for Tutorial 1*

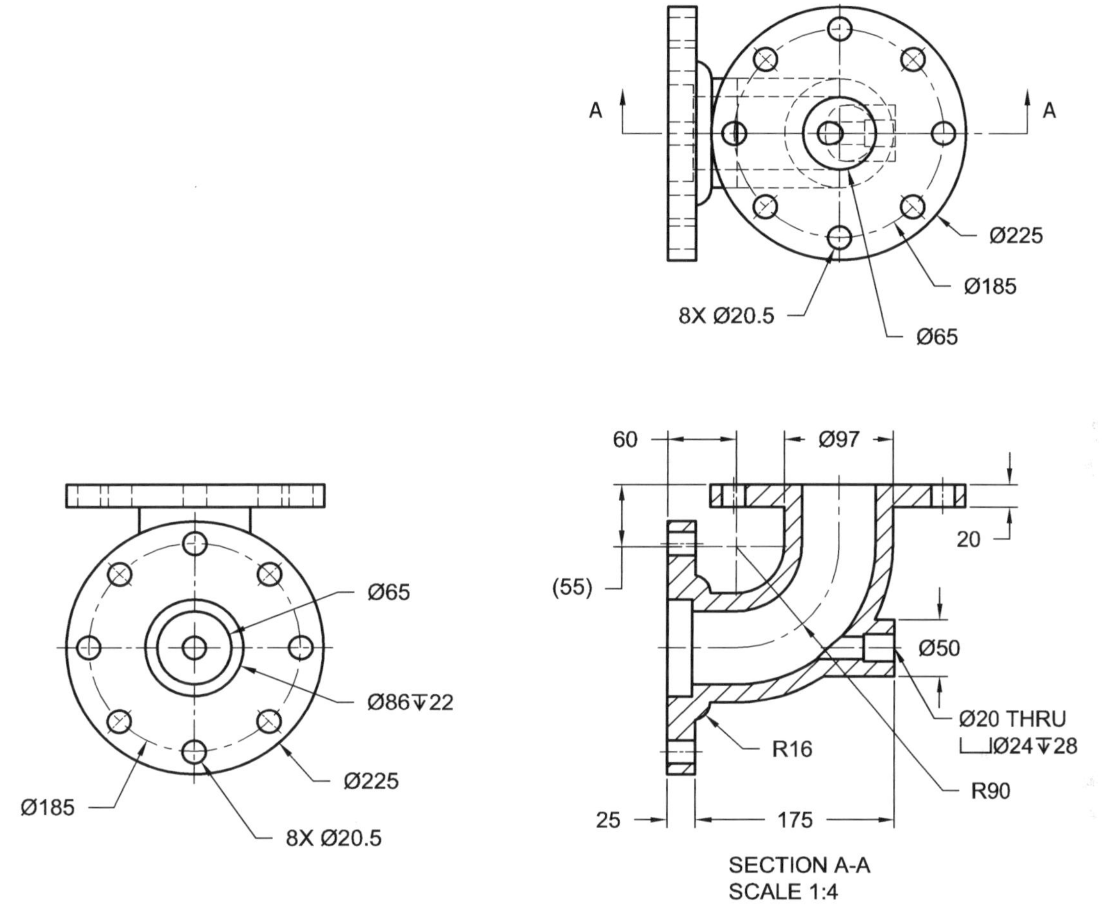

Figure 8-2 *Views and dimensions of the model for Tutorial 1*

Creating the Path for the Sweep Feature

As mentioned earlier, the base feature of the model is a sweep feature. To create the sweep feature, first you need to create its path on the XY plane. The path is a combination of two lines and an arc.

1. Start a new metric standard template file and create the path of the sweep feature on the XY plane from the origin. Add required dimensions. Exit the Sketching environment, and if required, change the view to the isometric view, as shown in Figure 8-3.

Creating the Work Plane Normal to the Start Section of the Path

After creating the path, you need to create a work plane normal to the start section of the path and position it at the start point of the path. A work plane is used to draw profile for a sweep feature. The start section of the path can be either a horizontal line or a vertical line of 35 mm length. In this tutorial, the horizontal line is taken as the start section of the path.

1. Choose the **Normal to Axis through Point** tool from **3D Model > Work Features > Plane** drop-down; you are prompted to select an edge/axis or a point.

2. Select the line that is at the bottom-left of the sketch and then click on its endpoint.

   As soon as you select the start point of the line, a work plane normal to the line is created and is positioned at the start point of the line. The work plane at the start point of the path is shown in Figure 8-4.

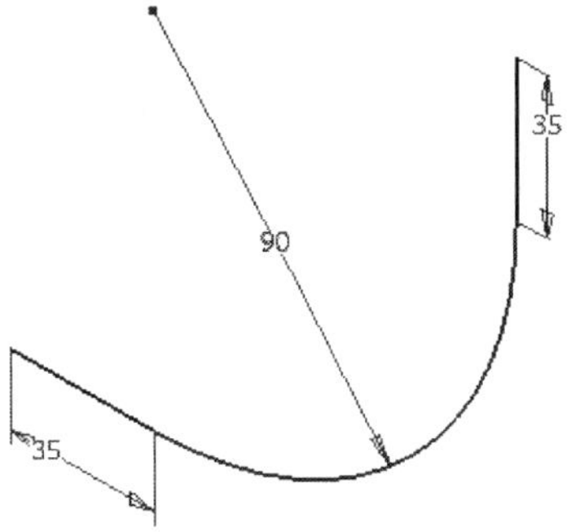

Figure 8-3 *Isometric view of the path for the sweep feature*

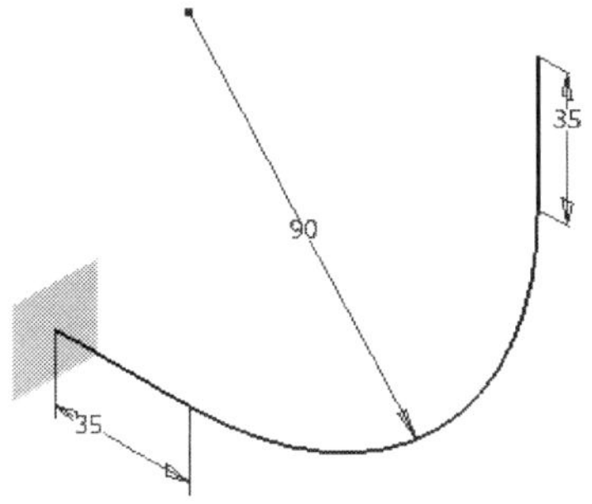

Figure 8-4 *Work plane normal to the path*

Drawing the Profile of the Sweep Feature

The profile of the sweep feature will be created on the new work plane. Therefore, you need to define a sketch plane on the new work plane.

1. Choose the **Start 2D Sketch** tool from **3D Model > Sketch** drop-down and then select the new work plane as the sketching plane.

   As soon as you select the new work plane as the plane for sketching, the Sketching environment is invoked. Notice that the origin of the Sketching environment coincides with the start point of the start section of the path. This helps you position the profile of the sweep feature.

2. Draw a circle of 97 mm diameter as the profile of the sweep feature. Take the center of the circle as the origin of the Sketching environment. Exit the Sketching environment and change the view to the isometric view, if required. The profile of the sweep feature is shown in Figure 8-5.

Sweeping the Profile

1. Choose the **Sweep** tool from the **Create** panel of the **3D Model** tab; the profile of the sweep feature is selected automatically in the graphics window and you are prompted to select the path.

2. Select the path and then choose **OK** from the **Properties-Sweep** dialog box. The sweep feature after changing the viewing direction is shown in Figure 8-6.

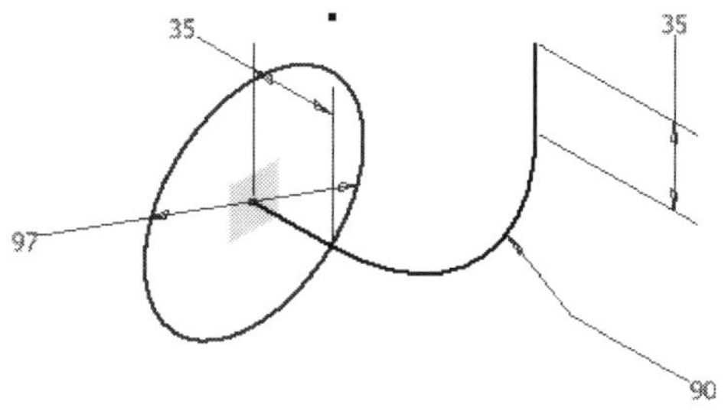

Figure 8-5 Profile of the sweep feature

Figure 8-6 The sweep feature

Creating the Shell Feature

The shell feature scoops out material from the sweep feature and leaves behind a model with some wall thickness. You need to remove the left and top faces of the sweep feature to view the cavity inside.

1. Choose the **Shell** tool from the **Modify** panel of the **3D Model** tab; the **Shell** dialog box is displayed and you are prompted to select the surfaces to be removed. In this dialog box, the **Inside** button is selected by default which is used to define the direction of shell creation which is inward of the model.

2. Select the left and top faces of the sweep feature; the selected faces are highlighted in blue. The diameter of the inner cavity is 65 mm and the diameter of the sweep feature is 97 mm. As a result, the wall thickness comes out to be 16 mm.

3. Enter **16** in the **Thickness** edit box and choose the **OK** button. The model after creating the shell feature is shown in Figure 8-7.

Tip

*An alternative way of creating the shell feature is by using the **Sweep** tool. In case of the **Sweep** tool, you need to create a path curve and two concentric circles of required size. When you sweep both circles along the path curve, the inner circle is subtracted from the outer one. This way the inner cavity can be created automatically. In this tutorial, you will use the **Shell** tool to create the inner cavity.*

Creating the Remaining Features

1. Create the remaining features by defining new sketch planes at the left face of the first feature that you created. Draw two concentric circles of diameter 225 mm and 65 mm. Exit the Sketching environment and then extrude area created between these two circles to a distance of 25 mm. Similarly, create the same feature on the top face of the first feature.

2. Select the back face of the feature that you created in the last step. Draw a circle of 129 mm diameter, also another concentric circle of 97 mm diameter and extrude it upto a depth of 16 mm in the required direction. Next, create a fillet of 16 mm on the outer edge of this feature.

   The join feature at the cylindrical tangent surface can be created by defining an offset work plane at a distance of 175 mm from the inner face, refer to Figure 8-2. You can create a hole on the join feature by using the **Hole** tool. Similarly, create a hole on the top and bottom features of the model and then create the pattern of these holes. The final model for Tutorial 1 is shown in Figure 8-8.

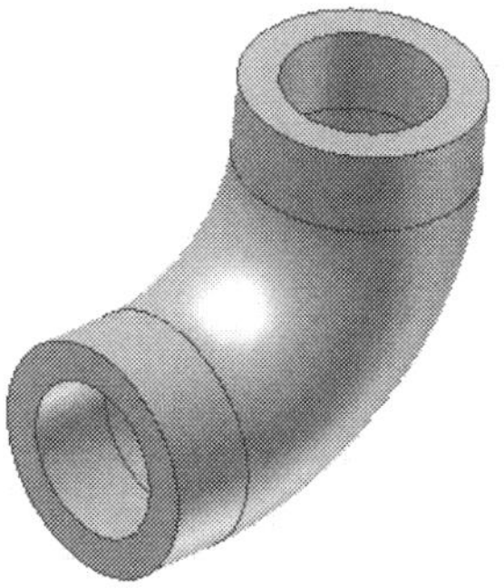

Figure 8-7 *Rotated view of model after creating the shell feature*

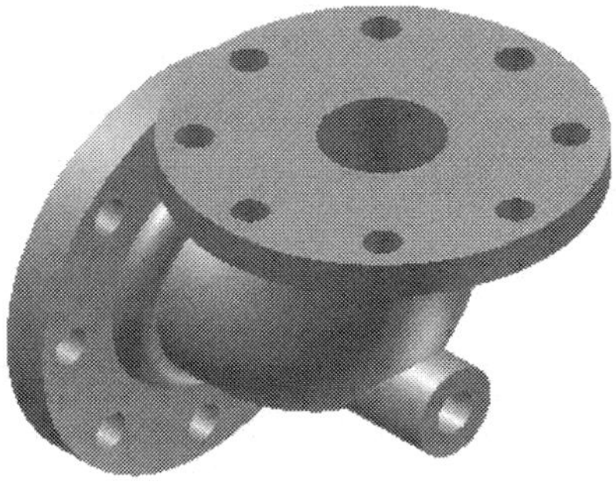

Figure 8-8 *Final model for Tutorial 1*

3. Save the model with the name *Tutorial1.ipt* at the location *C:\Inventor_2020\c08* and close the file.

Tutorial 2

In this tutorial, you will create the model of the Joint shown in Figure 8-9. Its dimensions are shown in Figures 8-10. The threads to be created are ANSI Metric M Profile of size 14 and designation M14x2. The class of the threads is 6g. Make sure the threads are right-handed. **(Expected time: 30 min)**

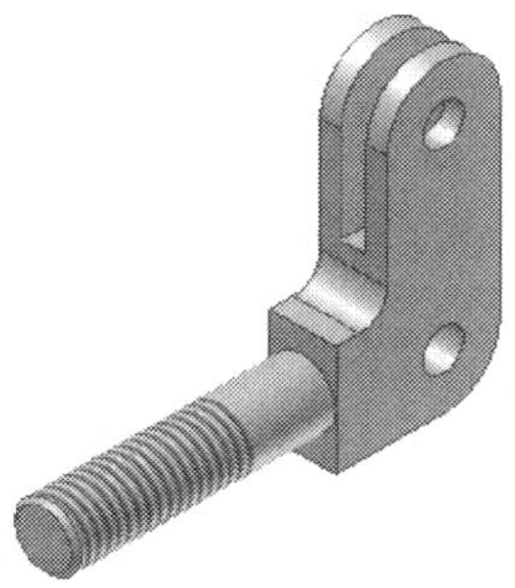

Figure 8-9 Solid model of the Joint

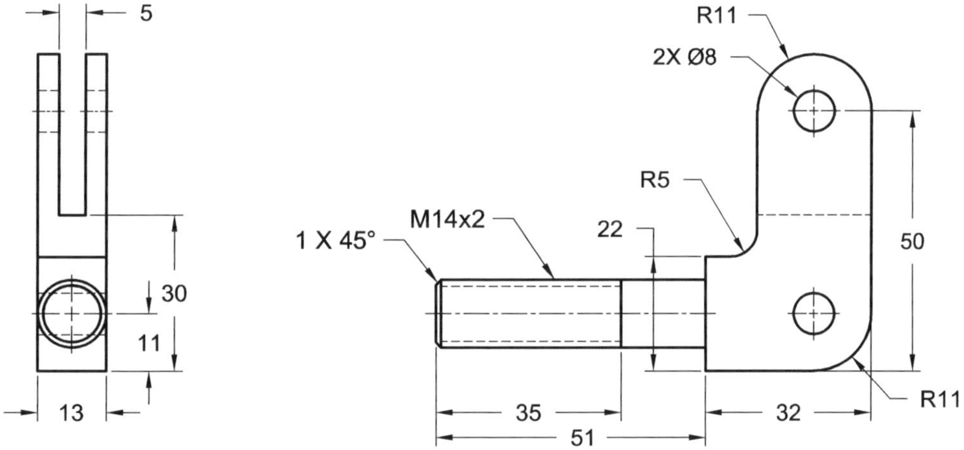

Figure 8-10 Dimensions of the Nut

The following steps are required to complete this tutorial:

a. Create the base feature of the model on the YZ plane.
b. Create the cut feature.
c. Create the cylindrical join feature on the left face and then create the chamfer feature.
d. Finally, create threads on the cylindrical join feature using the **Thread** tool.

Creating the Base Feature

1. Create the base feature of the model on the YZ plane, as shown in Figure 8-11. For dimensions of the base feature, refer to Figures 8-10.

Creating the Cut Feature in the Base Feature

1. Create the cut feature by defining a sketch plane on the right face of the base feature, as shown in Figure 8-12. For dimensions of the cut feature, refer to Figure 8-10.

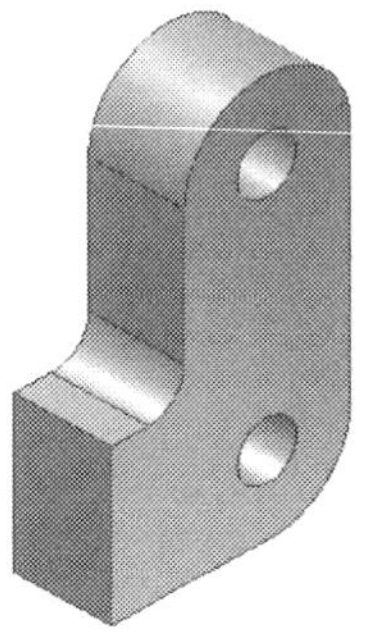

Figure 8-11 *Base feature for the model*

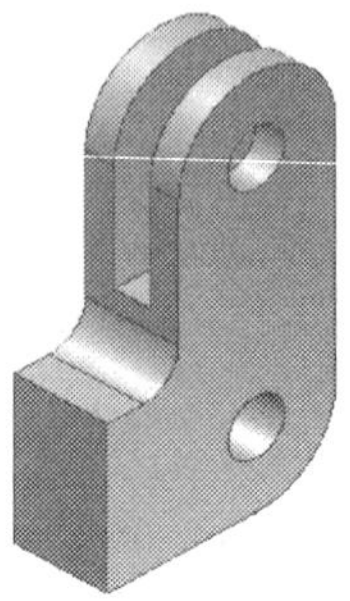

Figure 8-12 *Model after creating the cut feature*

Creating the Join and Chamfer Features

1. Create the cylindrical join feature, as shown in Figure 8-13. For dimensions, refer to Figures 8-10 .

2. Create the chamfer feature on the end face of the cylindrical feature using the **Distance and Angle** button of the **Chamfer** dialog box, refer to Figure 8-14. For dimensions, refer to Figure 8-10.

Figure 8-13 *Model after creating the join feature*

Figure 8-14 *Model after chamfering*

Creating Threads

1. Choose the **Thread** tool from the **Modify** panel of the **3D Model** tab to invoke the **Properties-Thread** dialog box.

   In the **Behavior** node, by default the **Full Depth** button is in active mode. As a result, thread will be created on the whole cylindrical join feature. To create a thread of specified length, you need to deactivate this button and specify depth in the **Depth** edit box.

2. In the **Behavior** node, click on the **Full Depth** button to deactivate it.

3. Click at a point on the cylindrical join feature from where you want to create the thread; an arrow indicating the direction of thread creation is displayed. Make sure that arrow points toward the required direction.

3. Enter **34** in the **Depth** edit box.

4. Select **ANSI Metric M Profile** from the **Type** drop-down list and **14** from the **Size** drop-down list in the **Thread** node.

5. Select **M14x2** from the **Designation** drop-down list and **6g** from the **Class** drop-down list. Make sure that the **Right hand** radio button is selected in the **Direction** area. Choose **OK** from the dialog box to create the threads. The model after creating threads is shown in Figure 8-15.

Figure 8-15 *Final model for Tutorial 2*

5. Save the model with the name *Tutorial2.ipt* at the location *C:\Inventor_2020\c08* and then close the file.

EXERCISES

Exercise 1

Create a solid model for Exercise 1, as shown in Figure 8-16. The dimensions to be used for creating the model are given in the same Figure. **(Expected time: 45 min)**

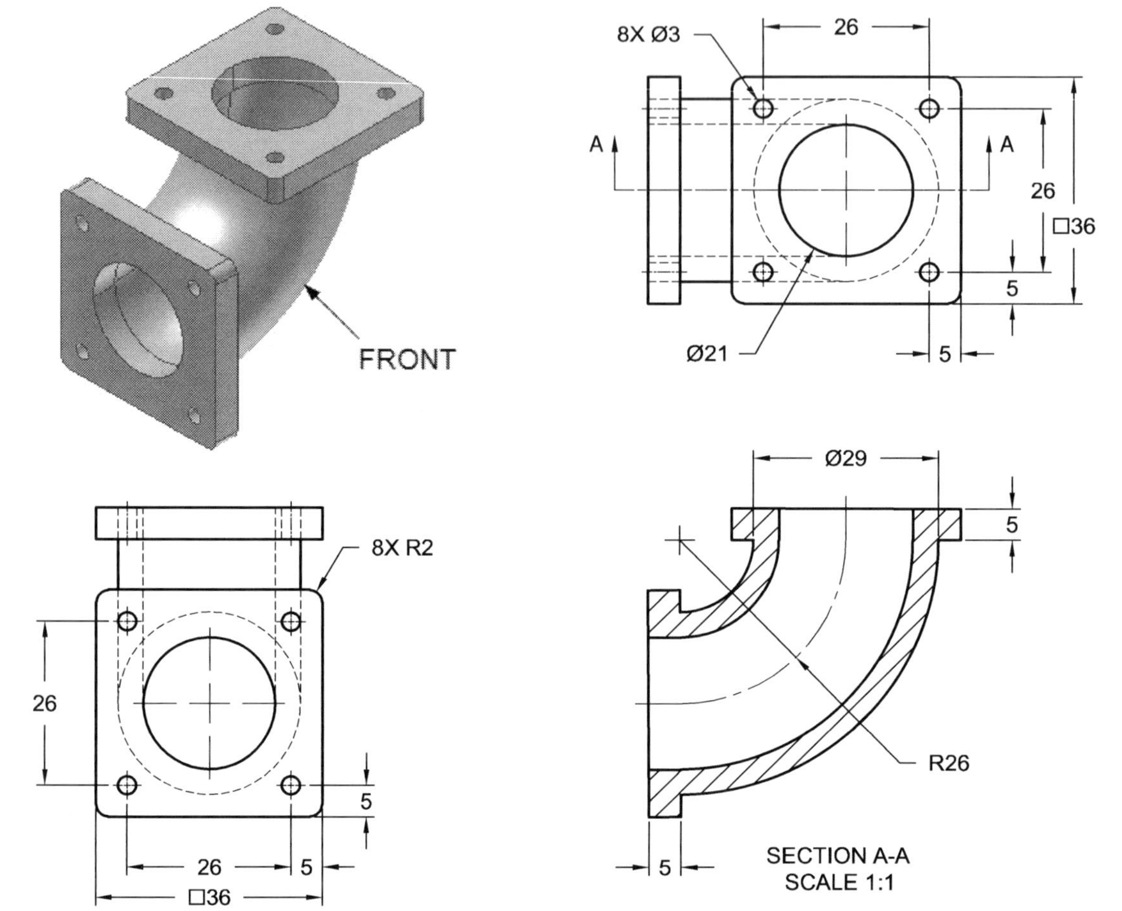

Figure 8-16 *Views and dimensions of the model*

Exercise 2

Create a solid model of the hexagonal Cap Screw shown in Figure 8-17. Its dimensions are shown in Figure 8-18. The threads to be created are ANSI Metric M Profile of size 10 and designation M10x1.5. The class of threads is 6g. Make sure the threads are right-handed.

(Expected time: 30 min)

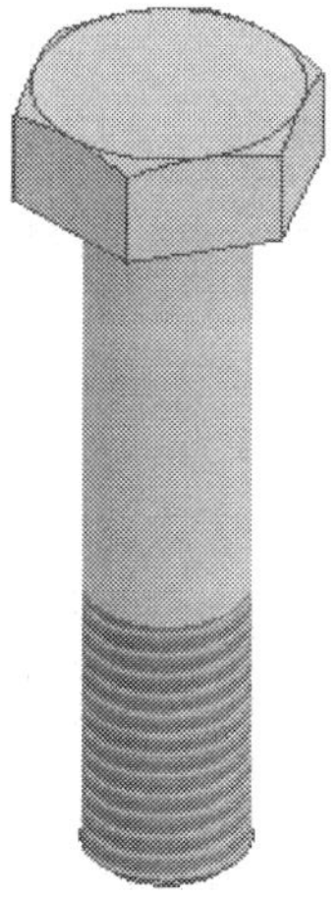

Figure 8-17 *Solid model of the Cap Screw*

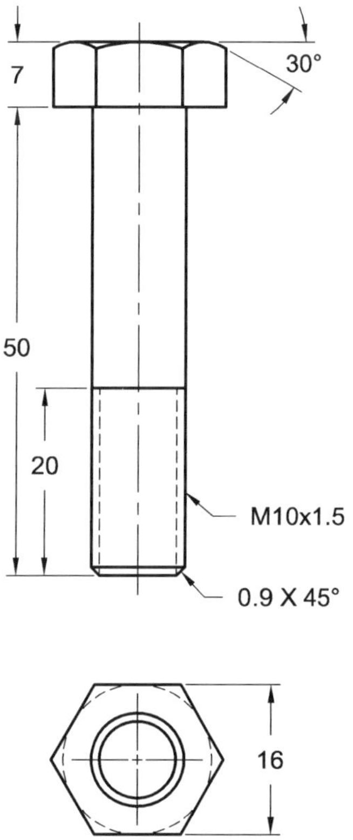

Figure 8-18 *Dimensions of the Cap Screw*

This page is intentionally left blank

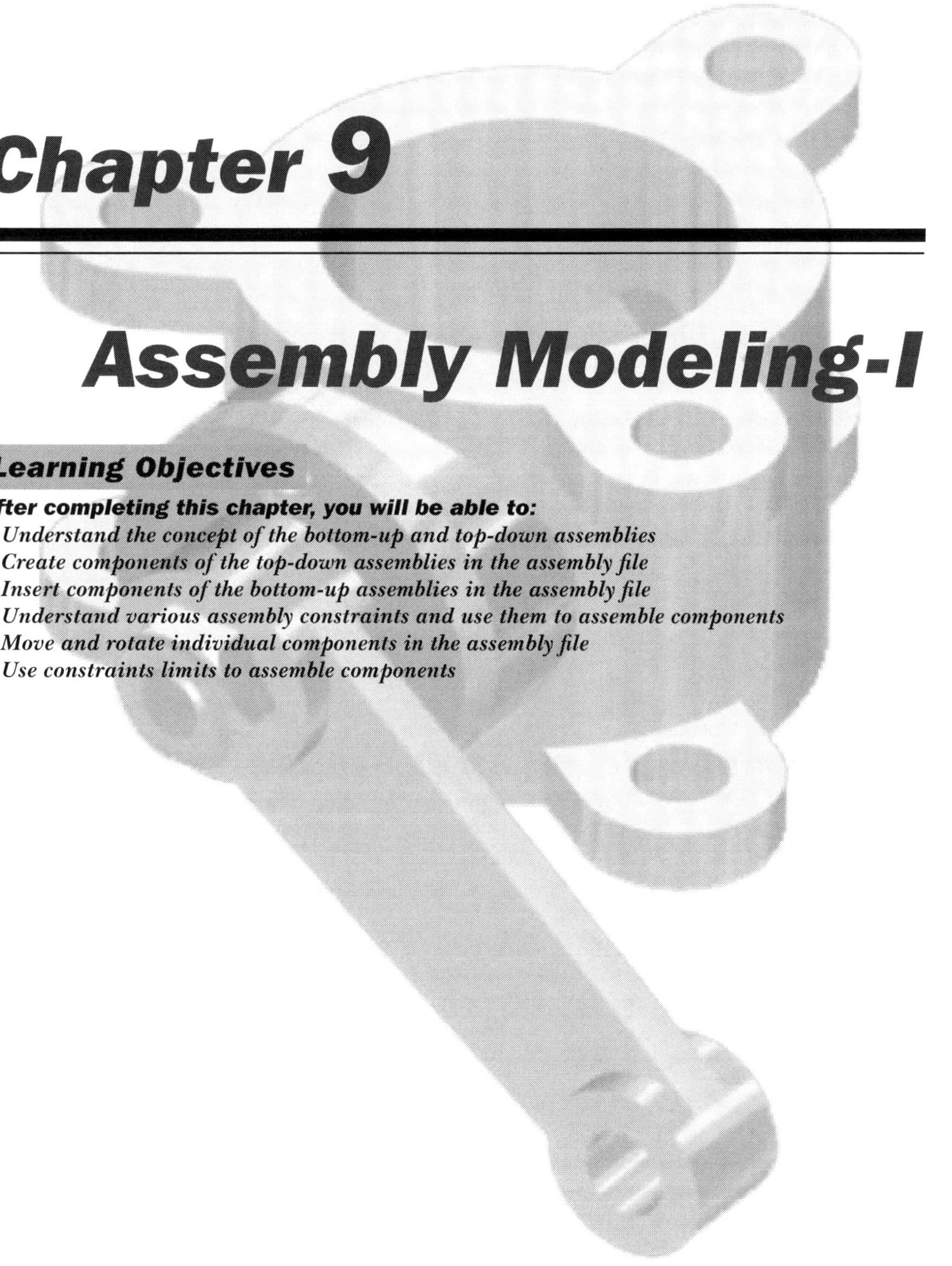

Chapter 9

Assembly Modeling-I

Learning Objectives

After completing this chapter, you will be able to:

- *Understand the concept of the bottom-up and top-down assemblies*
- *Create components of the top-down assemblies in the assembly file*
- *Insert components of the bottom-up assemblies in the assembly file*
- *Understand various assembly constraints and use them to assemble components*
- *Move and rotate individual components in the assembly file*
- *Use constraints limits to assemble components*

ASSEMBLY MODELING

An assembly design consists of two or more components assembled at their respective working positions. In Autodesk Inventor, the components of the assembly can be bound using the parametric assembly constraints. As the assembly constraints are parametric in nature, you can modify or delete them whenever you want. In Autodesk Inventor, the assemblies are created in the **Assembly** module.

TYPES OF ASSEMBLIES

In Autodesk Inventor, you can create two types of assemblies: top-down assemblies and bottom-up assemblies. Both these assemblies are discussed next.

Top-down Assemblies

A top-down assembly is an assembly whose components are created within the assembly file. In this type of assembly, first the components are created in the assembly file and then assembled using the assembly constraints.

Bottom-up Assemblies

A bottom-up assembly is an assembly whose components are created as separate part files and are referenced in the assembly file as external components.

CREATING TOP-DOWN ASSEMBLIES

As mentioned earlier, in top-down assemblies, all components are created within the assembly file. To create the components, you require the environment where you can draw the sketches of the sketched features and also the environment where you can convert the sketches into features.

CREATING BOTTOM-UP ASSEMBLIES

As mentioned earlier, in the bottom-up assemblies, the components are created as separate part files. All the individual part files are then inserted in an assembly file and are assembled using the assembly constraint. The first component inserted in the assembly will be grounded and its origin will coincide with that of the assembly file. Also, the three default planes of the part file will be placed in the same orientation as that of the default planes of the assembly file. The individual components are inserted in the assembly file using the **Place** tool.

ASSEMBLING COMPONENTS BY USING THE CONSTRAIN TOOL

Ribbon: Assemble > Relationships > Constrain

In Autodesk Inventor, the components are assembled using five types of assembly constraints, two types of motion constraints, and a transitional constraint. All these constraints are available in the **Place Constraint** dialog box that is displayed on choosing the **Constrain** tool. The dialog box consists of different tabs and each of them has one or more types of constraints.

USING ALT+DRAG TO APPLY ASSEMBLY CONSTRAINTS

Autodesk Inventor allows you to apply the assembly constraints without invoking the **Place Constraint** dialog box. This is done by pressing the ALT key and then dragging the component. The following steps explain the procedure to apply assembly constraints using the ALT+Drag method.

1. Press and hold the ALT key and then drag the required component toward the component with which it needs to be assembled; the symbol of the **Mate** constraint will be displayed below the cursor. This is because when you use the ALT+Drag method, by default, the **Mate** constraint is applied.

2. Release the ALT key but make sure you do not release the left mouse button. If you release the left mouse button, the constraints cannot be applied. Press SPACEBAR to change the mate position to the flush position. Drag the first component to the component that you want to select as the second component and then release the left mouse button; the assembly constraint will be applied.

TUTORIALS

Tutorial 1

In this tutorial, you will create the components for a Butterfly Valve assembly and then assemble them, refer to Figure 9-1. The Body and the Shaft will be created in the assembly file and the remaining components will be created as individual parts in separate part files. Therefore, you need to use a combination of top-down and bottom-up assembly approaches. The views and dimensions of the components are shown in Figures 9-2 through 9-9.

(Expected time: 3 hrs 30 min)

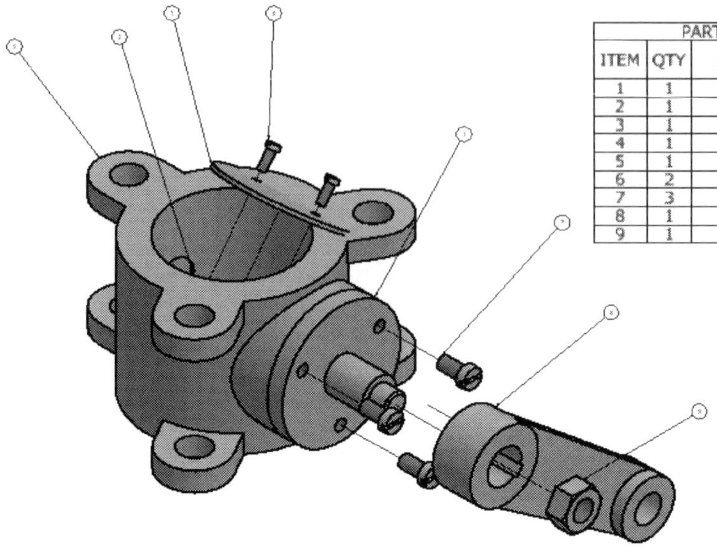

PARTS LIST		
ITEM	QTY	PART NUMBER
1	1	Body
2	1	shaft
3	1	Retainer
4	1	Arm1
5	1	Plate
6	2	scew1
7	3	Screw
8	1	Nut
9	1	Parallel key

Figure 9-1 *Exploded view of Butterfly Valve assembly*

Figure 9-2 *Solid model of the Body*

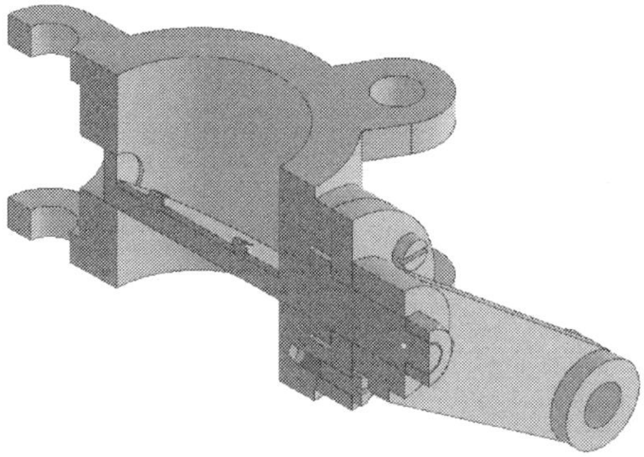

Figure 9-3 *Inside view of the Butterfly Valve assembly*

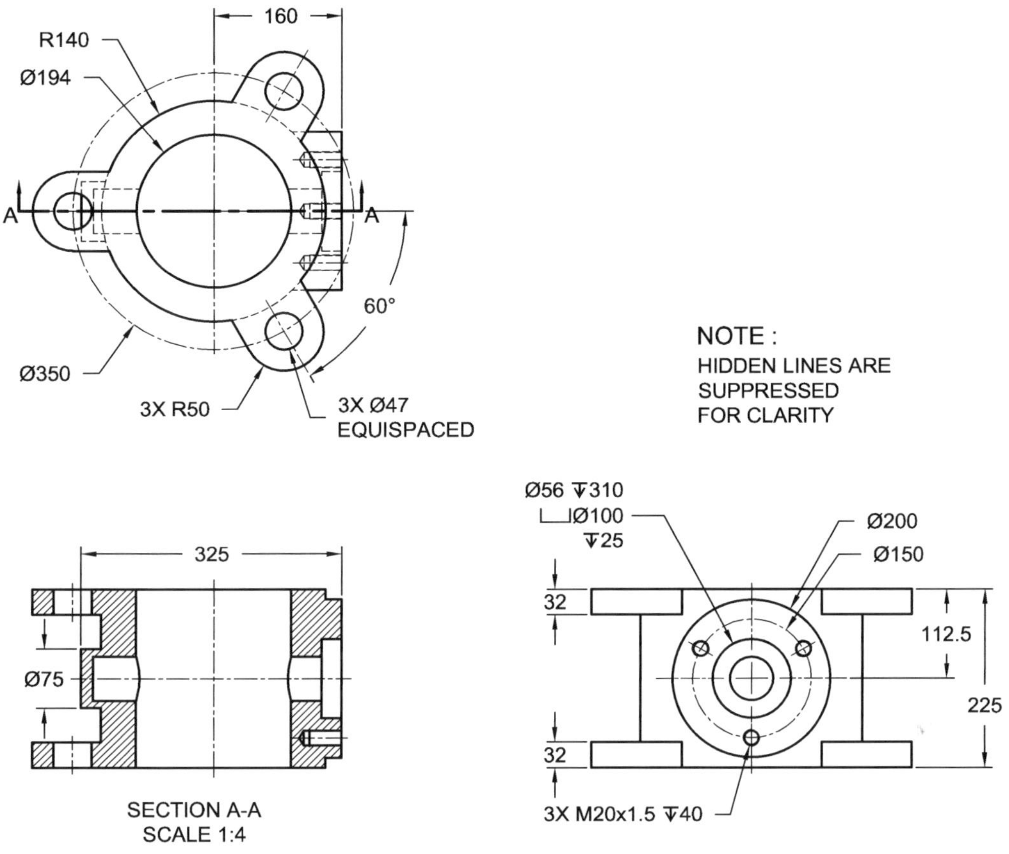

Figure 9-4 *Views and dimensions of the Body*

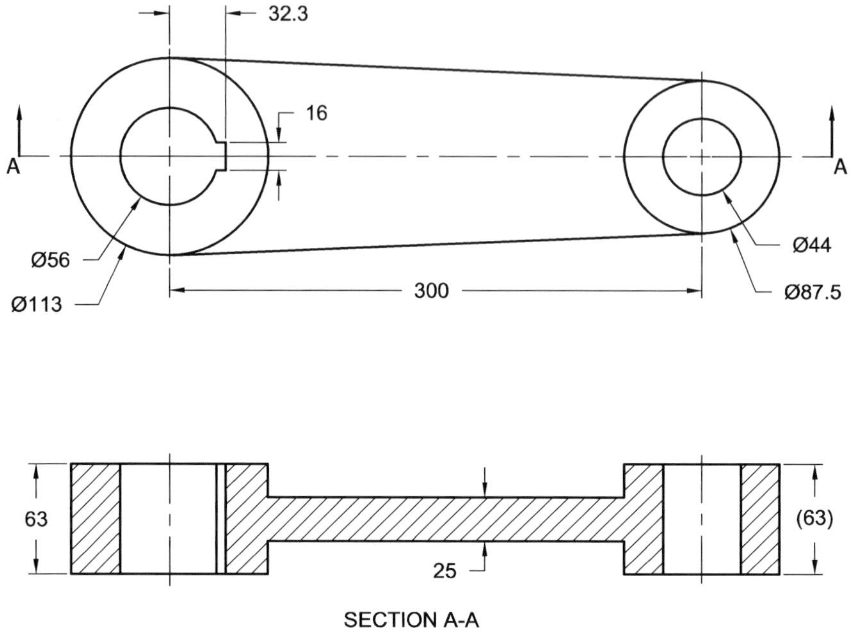

Figure 9-5 *Views and dimensions of the Arm*

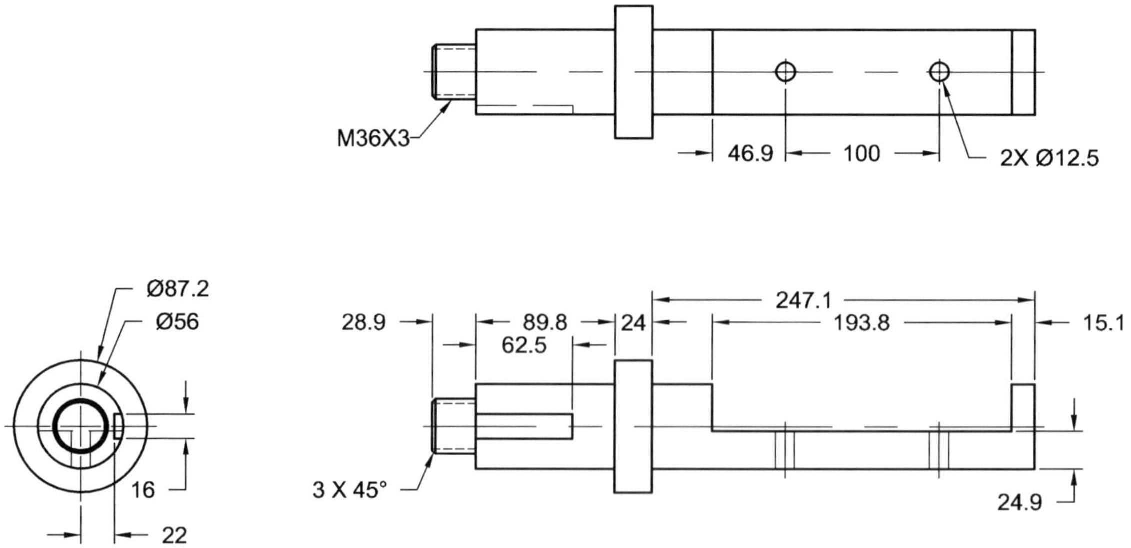

Figure 9-6 *Views and dimensions of the Shaft*

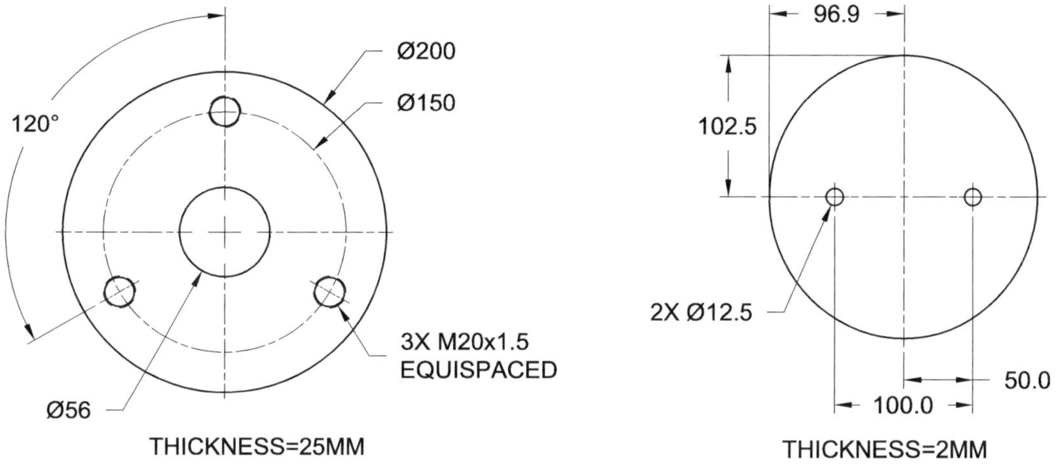

Figure 9-7 *Dimensions of the Retainer and Plate*

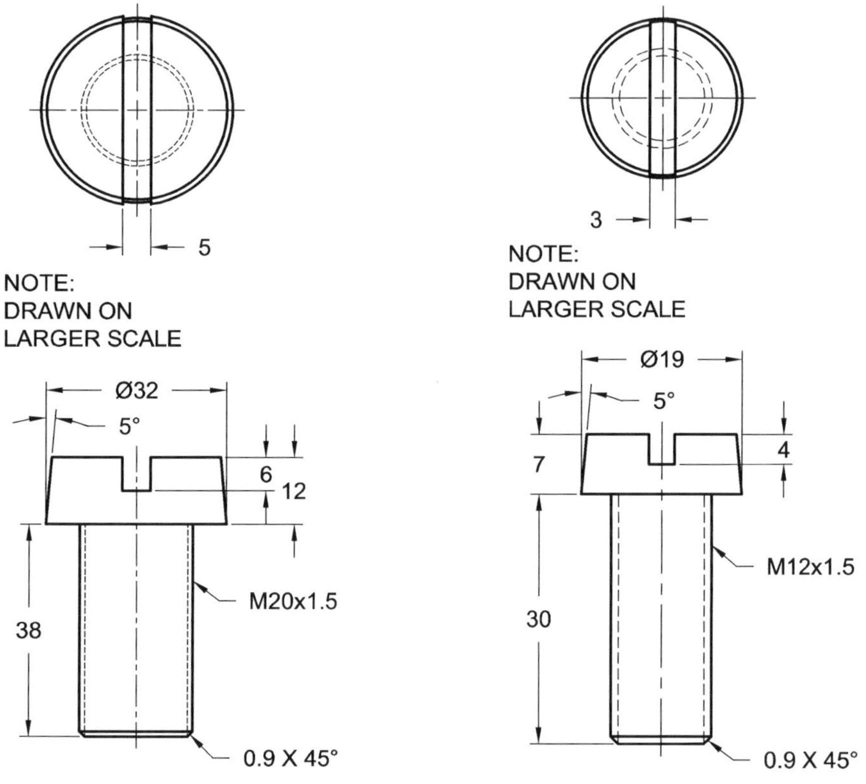

Figure 9-8 *Dimensions of the Screw and Screw1*

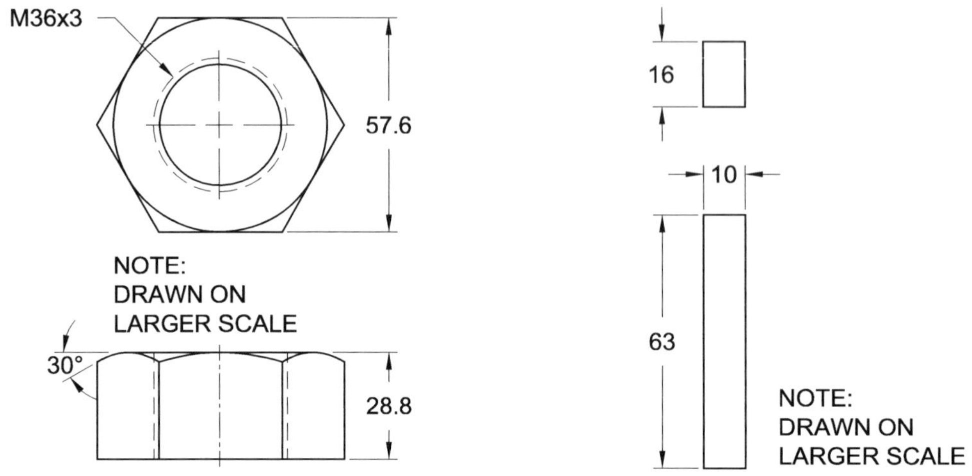

Figure 9-9 *Dimensions of the Nut and Parallel Key*

The following steps are required to complete this tutorial:

a. Start new metric standard part files and then create the other individual components.
b. Create the Body and the Shaft in the assembly file and then assemble these two components using the options in the **Place Constraint** dialog box. Save and close the assembly file.
c. Open the assembly file and insert the individual components in the assembly file using the **Place** tool.
d. Assemble the components using the **Place Constraint** dialog box to complete the Butterfly Valve assembly.

Creating a New Project for the Assembly

Before creating the new project file for the assembly, create a folder with the name *Butterfly Valve* at the location *C:\Inventor_2020\c09.*

1. Close all Autodesk Inventor files and then choose the **Projects** tool from the **Launch** panel of the **Get Started** tab; the **Projects** dialog box is displayed.

2. Choose the **New** button from the **Projects** dialog box; the **Inventor project wizard** dialog box is displayed.

3. Select the **New Single User Project** radio button if it is not selected by default and then choose the **Next** button from the **Inventor project wizard** dialog box.

4. Enter **Butterfly Valve** as the name of the new project in the **Name** edit box.

5. Choose the **Browse for project location** button available on the right of the **Project (Workspace) Folder** edit box; the **Browse for Folder** dialog box is displayed.

6. Browse to the location *C:\Inventor_2020\c09* and select the folder *Butterfly Valve*. Next, choose the **OK** button from the **Browse for Folder** dialog box.

7. Choose the **Next** button and then the **Finish** button from the **Inventor project wizard** dialog box to exit it.

8. Double-click on the newly added project in the **Project name** area to make it current, if it is not selected by default and then choose the **Done** button to exit the **Projects** dialog box.

Creating the Body

You need to create the Body and the Shaft in the assembly file by using the top-down approach of assembly modeling. To create these two components, you first need to start a new metric assembly file.

1. Choose the **New** tool from the **Quick Access Toolbar** to invoke the **Create New File** dialog box. Next, choose the **Metric** tab from this dialog box and double-click on **Standard (mm).iam** to start a metric assembly file. The assembly environment is invoked.

   You will notice that only a few tools are enabled in the **Assemble** tab. This is because no component is present in the assembly file. Once a component is placed or created, all other tools will be available for use.

2. Choose the **Create** tool from the **Component** panel of the **Assemble** tab.

3. Enter **Body** as the name of the new part file in the **New Component Name** edit box.

4. Choose the **Browse Templates** button; the **Open Template** dialog box is displayed. Select **Standard (mm).ipt** from the **Metric** tab of this dialog box. Choose the **OK** button from the **Open Template** dialog box to exit.

5. Specify the location of the new part file in the **New File Location** edit box as *C:\Inventor_2020\c09\Butterfly Valve*.

6. Choose the **OK** button from the **Create In-Place Component** dialog box; you are prompted to specify a sketching plane for the base feature. Select the **XY Plane** of the main assembly from the **Browser Bar**.

7. Invoke the sketching environment and select the **Start 2D Sketch** button from the **Sketch** Panel of the **3D Model** tab. Now, choose the **Home** button of the ViewCube to get the right orientation.

8. Next, select the **XY Plane** of the Body.

Tip

It is recommended that you create separate folders for saving individual component files of assemblies because a number of assemblies have components with similar names. For example, the name Body is commonly used for a number of assemblies. Therefore, if you create a part, name it as Body and then store it in the folder of a particular assembly, so that there is no confusion in placing the components. Also, when you open the assembly next time, there will be no confusion in referring to the required component.

Note

Remember that if you save the file when the part modeling environment is active, then only the part file will be saved and not the assembly file. This means while creating the Body, if you choose the ***Save*** *tool from the* ***Quick Access Toolbar****, the Body.ipt file will be saved and not the current assembly file. To save the current assembly file, you need to exit the part modeling environment and then choose the* ***Save*** *tool in the assembly modeling environment.*

9. Create the **Body** of the Butterfly Valve using the dimensions given in Figures 9-4. The assembly file after creating the **Body** is shown in Figure 9-10.

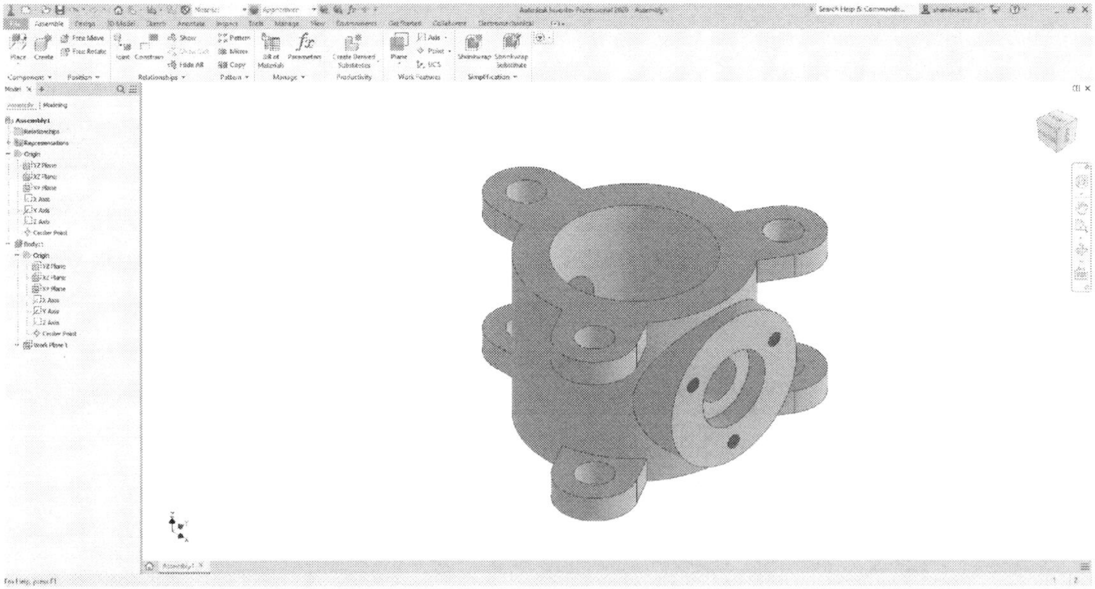

Figure 9-10 *Assembly file in the part modeling environment after creating the Body*

You will notice that the part modeling environment is still active in the assembly file. To proceed further, you need to save the part file and then exit the part modeling environment.

10. Choose the **Save** tool to save the part file and then choose the **Return** tool from the **Return** panel of the **3D Model** tab to exit the part modeling environment.

When you choose the **Return** tool, you will notice that the **Assemble** tab is chosen in place of the **3D Model** tab.

As mentioned earlier, if you choose the **Save** tool before exiting the part modeling environment, only the part file will be saved. The assembly file will be saved only after you exit the part modeling environment.

11. Choose the **Save** tool from the **Quick Access Toolbar** and save the assembly with the name *Butterfly Valve* in the *Butterfly Valve* folder.

Creating the Shaft

The second component that has to be created in the assembly file is the Shaft. Therefore, you need to again activate the part modeling environment and the sketching environment to create the Shaft. But, as the Body is already present in the assembly file, it might restrict the view of the part that you will be creating next. Considering this, the part modeling environment is designed in such a way that when you start creating the components in the assembly file, all existing components become transparent and the view of the newly created parts is not restricted.

1. Choose the **Create** tool from the **Component** panel of the **Assemble** tab or choose the **Create Component** option from Marking menu to invoke the **Create In-Place Component** dialog box.

2. Enter the name of the new part file as **Shaft** in the **New Component Name** edit box of the **Create In-Place Component** dialog box.

3. Choose the **Browse Templates** button on the right of the **Template** drop-down list to invoke the **Open Template** dialog box. In this dialog box, choose **Metric** and then open the **Standard (mm).ipt** template.

4. Specify the location of the new part file in the **New File Location** edit box as *C:\Inventor_2020\c09\Butterfly Valve*.

5. Clear the **Constrain sketch plane to selected face or plane** check box and then choose **OK**; you are prompted to select the plane for the base feature.

6. Select **XY Plane** of the main assembly from the **Browser Bar**.

7. Invoke the sketching environment by choosing the **Start 2D Sketch** tool from the **Sketch** panel of the **3D Model** tab. Now choose the **Home** button of the ViewCube to get the right orientation and then select **XY Plane** of the Shaft from the **Browser Bar**; the Body becomes transparent. You can now start creating the Shaft.

8. Create the Shaft and then save it. Next, exit the part modeling environment by choosing the **Return** tool from the **Return** panel of the **3D Model** tab. Save the assembly file by choosing the **Save** tool from the **Quick Access Toolbar**.

   When you exit the part modeling environment, you will notice that the Body is no more transparent. Also, both components in the assembly file interfere with each other. Therefore, before proceeding with assembling of these components, you need to move one of the components such that it does not interfere with the other. You can move the individual component by using the **Free Move** tool.

9. Choose the **Free Move** tool from the **Position** panel of the **Assemble** tab and then move the cursor over the Body; you are prompted to drag the component to a new location. Select the Body and drag it to a new location where it does not interfere with the Shaft. Choose the **Zoom All** tool to increase the display area. The assembly file with both the components is shown in Figure 9-11. 

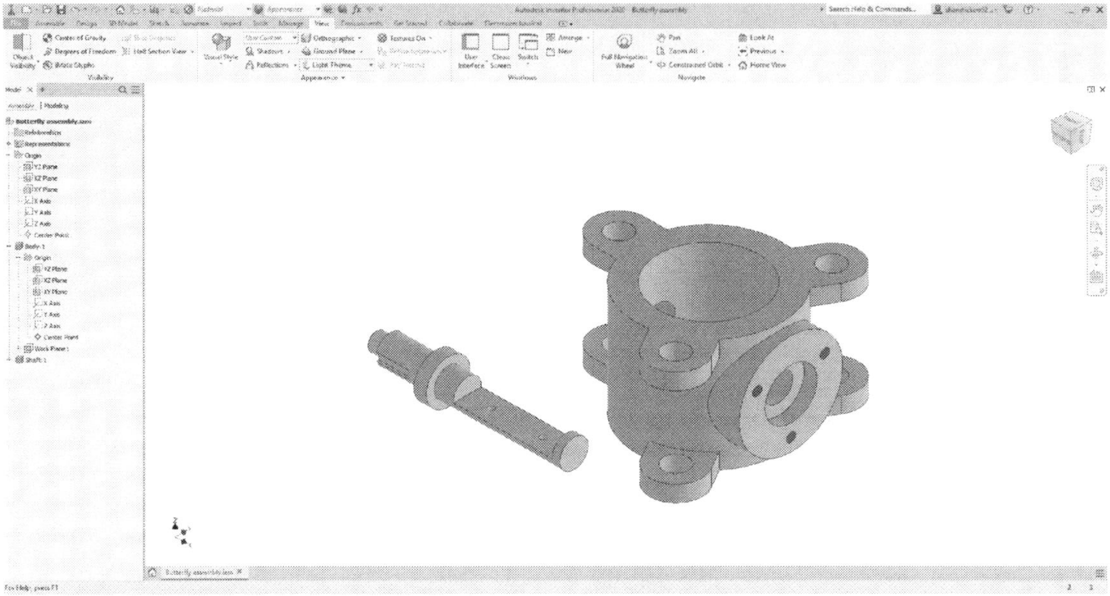

Figure 9-11 *The assembly file after creating the Body and the Shaft*

Note

If the orientation of the Shaft and the Body on your computer screen is different from the one shown in Figure 9-11, you can reorient them using the ***Free Rotate*** *tool from the Marking menu.*

Assembling the Components

The Shaft has to be inserted in the counterbore hole of the Body. Therefore, you can use the **Insert** constraint to assemble these components. As mentioned earlier, the **Insert** constraint forces the selected components or features to share the same location and orientation of the central axis. It also makes the selected faces coplanar. Therefore, the Shaft will be assembled with the Body using the **Insert** constraint.

1. Choose the **Constrain** tool from the **Relationships** panel of the **Assemble** tab or choose **Constraint** from the Marking menu to invoke the **Place Constraint** dialog box.

   By default, the **Mate** constraint is selected. But you need the **Insert** constraint for assembling the Shaft with the Body.

2. Choose the **Insert** button from the **Type** area of the **Assembly** tab in the **Place Constraint** dialog box; you will notice that the **Insert** constraint symbol is attached to the cursor. This symbol moves along with the cursor when you move the cursor in the graphics window. 

3. Select the first edge on the Shaft, as shown in Figure 9-12.

   You will notice that the selected edge is highlighted and an arrow is displayed along the direction of the central axis of the Shaft. This arrow will also point in the direction in which the Shaft will be assembled. Also, the **2 (Second Selection)** button in the **Selections** area of the **Place Constraint** dialog box is automatically chosen. Choose the **Opposed** option from the **Solution** area.

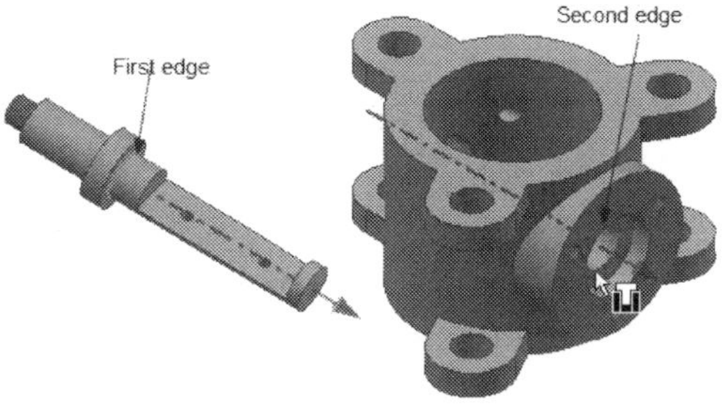

Figure 9-12 *Selecting the edges to apply the* ***Insert*** *constraint*

4. Select the inner edge of the counterbore hole as the second edge, refer to Figure 9-12. As soon as you select the second edge for applying the constraint, the preview of the Shaft assembled with the Body is displayed. This is because the **Show Preview** check box is selected by default in the **Place Constraint** dialog box.

5. Choose the **Apply** button to assemble the Shaft with the Body and then choose **Cancel** from the dialog box to exit. The Body and the Shaft will be constrained together.

Creating Other Components

1. Save the current assembly file and then close it by choosing **Close > Close** from the **Application Menu**.

2. Create the other components as individual part files and save them with their names in the *Butterfly Valve* folder.

3. Exit the part files and then again open the *Butterfly Valve.iam* file.

Assembling the Retainer

The next component to be assembled is the Retainer. The Retainer is a circular part and so it can be assembled using the **Insert** constraint. The three holes of the Retainer have to match those on the front planar face of the Body. Also, the central hole of the Retainer has to match with the central hole of the front planar face of the Body. Therefore, you need to apply the **Insert** constraint twice - first time to align one of the smaller holes on the Retainer with one of the smaller holes on the front flat face of the Body, and second time to align the central holes. But first you need to place the Retainer in the assembly using the **Place** tool.

1. Choose the **Place** tool from the **Component** panel of the **Assemble** tab; the **Place Component** dialog box is invoked.

2. Select Retainer and then choose the **Open** button; the **Open** dialog box is closed and the Retainer is attached to the cursor. Also, you are prompted to place the component.

3. Place the Retainer at a location where it does not interfere with the existing components. After you have placed an instance of the Retainer, you are again prompted to place the component. As you need to place only one instance of the Retainer, you can exit the component placement option.

4. Right-click in the graphics window and choose **OK** from the Marking menu to exit the component placement option.

5. Choose the **Constrain** tool from the **Relationships** panel of the **Assemble** tab; the **Place Constraint** dialog box is displayed. If the **Place Constraint** dialog box is restricting the viewing of the components in the graphics window, you can move it by selecting its title bar and dragging it.

6. Choose the **Insert** button from the **Type** area. Select the circular edge of one of the smaller holes on the top face of the Retainer as the first edge, see Figure 9-13.

7. Select the circular edge of one of the smaller holes on the front planar face of the circular feature on the Body to apply the constraint, see Figure 9-13. Next, choose the **Apply** button.

   As soon as you select the second edge, the Retainer moves from its location and is assembled with the Body such that both the selected holes are concentric and the top face of the Retainer is coplanar with the front planar face of the circular feature on the Body. However, you will notice that the central hole of the Retainer is not concentric with the central hole of the left circular feature of the Body and the Shaft. Therefore, you need to apply the **Insert** constraint once again to align these components.

8. Select the circular edge of the Retainer that is coplanar with the Body as the first edge to apply the constraint, see Figure 9-14. You may have to rotate the model to select this edge.

9. Select the circular edge of the front circular feature on the Body as the second face to apply the constraint, see Figure 9-14. Choose **Apply** to assemble the components and then choose **Cancel** to exit the dialog box.

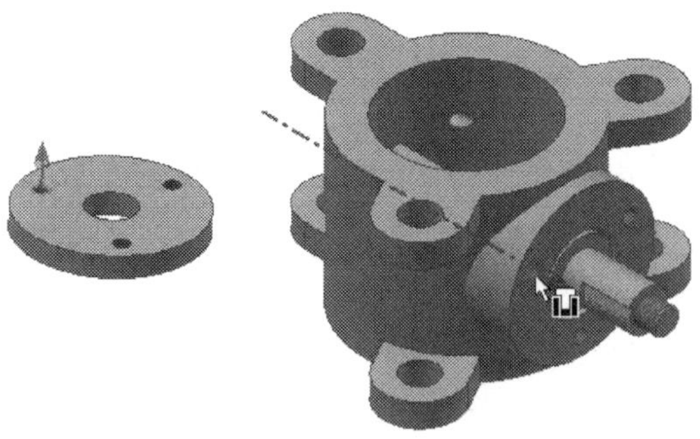

Figure 9-13 *Selecting the edges to apply the* ***Insert*** *constraint*

Figure 9-14 *Selecting the edges to apply the* ***Insert*** *constraint again*

Assembling the Arm

The next component to be assembled is the Arm. You need to use two constraints to assemble it. The first constraint is the **Insert** constraint and the second constraint is the **Angle** constraint which will be used to apply an angle between the XZ plane of the Arm and the top face of the Body. You will place the Arm using the **Place** tool.

1. Choose the **Place** tool from the **Component** panel of the **Assemble** tab to invoke the **Place Component** dialog box.

2. Double-click on the Arm; the Arm gets attached to the cursor.

3. Place the Arm at a location where it does not interfere with the existing components.

4. Right-click in the graphics window and choose **OK** from the Marking menu to exit the component placement option.

5. Choose the **Constrain** tool from the **Relationships** panel of the **Assemble** tab or from the Marking menu; the **Place Constraint** dialog box is displayed.

6. Choose the **Insert** button and then select the top circular edge of the hole with the keyway in the Arm as the first edge, as shown in Figure 9-15.

7. Select the circular edge on the front planar face of the Retainer as the second edge to apply the **Insert** constraint, refer to Figure 9-15. Next, choose the **OK** button.

   After performing these steps, the Arm will be assembled with the Retainer and the Shaft will be inserted in the hole with the key way of the Arm. The second constraint will be used to reorient the Arm such that it is assembled at an angle to the top face of the Body. This angle is the same as the angle between the top face of the Body and the flat face of the Shaft.

   Figure 9-16 shows the assembly after the Arm was assembled.

8. Now, it is important to apply the constraint between the Shaft and the Arm in such a manner that the Arm rotates with the Shaft. To do so, invoke the **Constrain** tool from the Marking menu; the **Place Constraint** dialog box is displayed.

9. Rotate the Arm in such a manner that key hole of Shaft and Arm match approximately.

10. Next, choose the **Angle** button from the **Type** area of the **Assembly** tab in the **Place Constraint** dialog box.

11. Select XY plane of the Shaft. Next, select the YZ plane of the Arm, as shown in Figure 9-17. Note that in this case, it is presumed that the sketch of the shaft is created by using the revolve feature on XZ plane. Therefore, XZ plane is passed through the center of Shaft.

Figure 9-15 *Selecting the edges to apply the* ***Insert*** *constraint*

Figure 9-16 *Assembly after assembling the Arm*

12. Choose the **Undirected Angle** button from the **Solution** area of the **Place Constraint** dialog box and enter **0** in the **Angle** edit box. Next, choose the **OK** button in the **Place Constraint** dialog box.

    The assembly after applying the constraint is shown in Figure 9-18.

Figure 9-17 *Selection of planes of Shaft and Arm for applying the* ***Angle*** *constraint*

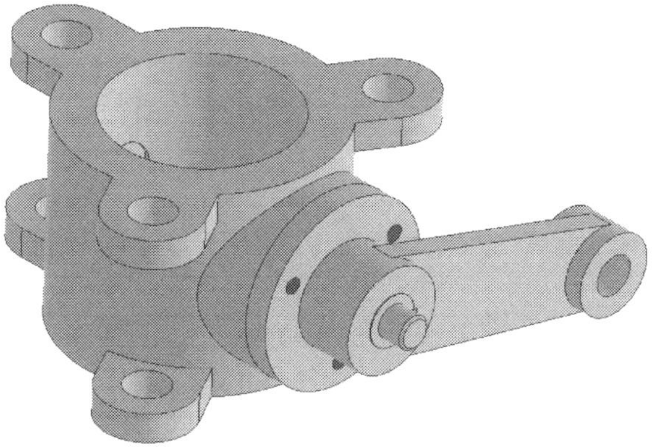

Figure 9-18 *Assembly after applying the constraint between the Shaft and the Arm*

13. Next, choose the **Constrain** tool from the Ribbon to display the **Place Constraint** dialog box again.

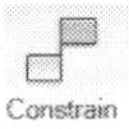

14. Choose the **Angle** button from the **Type** area in this dialog box; the symbol of the **Angle** constraint icon is attached to the cursor, suggesting that the process of assembling the components is resumed.

15. Select the XY plane of the shaft to apply the **Angle** constraint, refer to Figure 9-19.

16. Select the top face of the Body as the second face to apply the **Angle** constraint, refer to Figure 9-19.

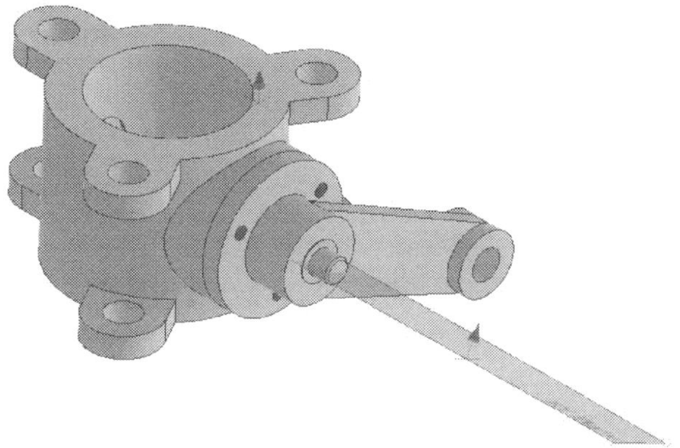

Figure 9-19 *Selecting faces to apply the* ***Angle*** *constraint*

17. Next, choose the **Undirected Angle** button from the **Solution** area and then choose the **More** button.

18. Select the **Maximum** check box in the **Limits** area and clear the **Use Angle As Resting Position** check box, if it is not cleared.

19. Enter **135** in the **Maximum** edit box and **0** in the **Minimum** edit box. Next, choose **Apply** and then **Cancel** from the **Place Constraint** dialog box to apply the constraint and exit the dialog box.

    The **Angle** constraint is applied between the components of the assembly.

Assembling the Plate

The next component to be assembled is the Plate. You first need to place the Plate in the assembly file and then assemble it on the flat face of the Shaft. You need to apply the **Insert** constraint twice to assemble the Plate with the Shaft. The first application of the constraint will align one of the holes on the Plate with one of the holes on the Shaft. The second application of the constraint will align the second hole on the Plate with a hole on the Shaft.

Since the Shaft is assembled inside the Body, the Body will restrict viewing of the components being assembled. To avoid this, Autodesk Inventor allows you to turn off the display of the components that you do not require for assembling the other components. Therefore, before proceeding with assembling of the Plate, you can turn off the display of the Body. This is done using the **Browser Bar**.

1. Right-click on the Body in the **Browser Bar** to display the shortcut menu. You will notice that a tick mark is displayed in front of the **Visibility** option in the shortcut menu. This

suggests that the display of this component is turned on. Choose the **Visibility** option again to turn off the display of the Body.

Note that the component whose visibility is turned off will be displayed in gray color in the **Browser Bar**.

2. Choose the **Place** tool from the **Component** panel of the **Assemble** tab to invoke the **Place Component** dialog box. While placing the component if the orientation of Plate is not same as shown in Figure 9-20, right-click in the graphics window and choose the Rotate X 90 degree option from the Marking menu.

3. Double-click on the Plate; the Plate gets attached to the cursor.
4. Place the Plate at a location where it does not interfere with the existing components.
5. Right-click in the drawing window and choose **OK** from the Marking menu to exit the component placement option.
6. Choose the **Constrain** tool from the **Relationships** panel of the **Assemble** tab; the **Place Constraint** dialog box is displayed.
7. Choose the **Insert** button and then select the circular edge of one of the holes on the top face of the Plate as the first edge to apply the constraint, refer to Figure 9-20.

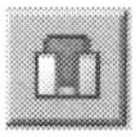

8. Select the circular edge of the right hole on the flat face of the Shaft as the second edge to apply the constraint, see Figure 9-20. Then, choose the **Apply** button to apply the constraint. As soon as you select the second face to apply the constraint, the Plate will move from its location and will be assembled with the Shaft. Now, the second constraint has to be applied to the other hole of the Plate. To do so, you need to select the face that is made coplanar with the flat face of the Shaft. Therefore, you need to reorient the model such that the back face of the Plate is visible and you can select the hole on that face to apply the constraint.
9. Rotate the assembly using the ViewCube such that the back face of the Plate is visible.
10. Select the circular edge on one of the holes on the back face of the Plate as the first edge to apply the constraint.

    Since you have rotated the model such that the back face of the Plate is visible, the flat face of the Shaft is not visible in the current view. Therefore, you need to switch back to the previous view. Sometimes when you use the **Place** tool, you cannot use the F5 key to invoke the previous view. In such cases, you need to invoke the **Free Rotate** tool and then right-click to display a shortcut menu. Then, you need to choose **Previous View** from this menu to switch back.

Figure 9-20 *Selecting the faces to apply the constraint*

11. Press the F5 key or choose the **Free Rotate** tool from the **Position** panel and then right-click in the drawing window to display a shortcut menu. Choose the **Previous View** option to switch back to the previous view.

12. Select the other hole on the flat face of the Shaft to apply the constraint. Choose the **Apply** button to apply the constraint and then choose the **Cancel** button to exit the dialog box. The assembly after assembling the Plate is shown in Figure 9-21.

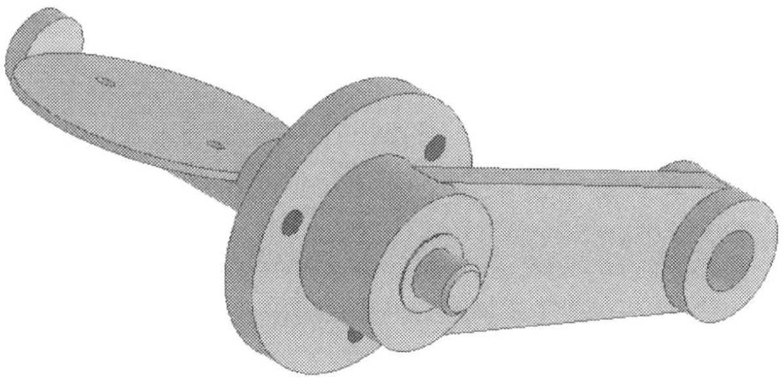

Figure 9-21 *Assembly after assembling the Plate*

13. Now, assemble the two instances of Screw 1 with two smaller holes on Plate by applying Insert constraint.

14. Turn on the visibility of the Body by using the **Browser Bar**.

Assembling the Screws

There are three instances of the Screw that need to be assembled such that they are inserted into three holes on the Retainer.

1. Turn off the display of the Arm by using the **Browser Bar**.

2. Choose the **Place** tool from the **Component** panel to invoke the **Place Component** dialog box.

3. Double-click on the Screw; the Screw gets attached to the cursor.

4. Place three instances of the Screw at a location where they do not interfere with the existing components.

5. Choose the **Constrain** tool from the **Relationship**s panel of the **Assemble** tab; the **Place Constraint** dialog box is displayed.

6. Choose the **Insert** button and then select the circular edge on the flat face of the head of the Screw as the first face to apply the constraint.

7. Select the circular edge of one of the smaller holes on the front face of the Retainer as the second face to apply the constraint. Next, choose the **Apply** button to apply the constraint.

8. Similarly, assemble the other two Screws with the other two smaller holes on the Retainer.

9. Turn on the visibility of the Arm by using the **Browser Bar**.

Assembling the Parallel Key

Next, you need to assemble the Parallel key with the Shaft and the Arm to prevent the relative rotation.

1. Choose the **Place** tool from the **Component** panel to invoke the **Place Component** dialog box.

2. Double-click on the Parallel key; the Parallel key gets attached to the cursor.

3. Place the Parallel key to the location where this could not interfere with the existing assembly. Next, rotate the Parallel key using the **Free Rotate** tool such that the flat face of the Parallel key is visible in the graphics window.

4. Choose the **Constrain** tool from the **Relationship**s panel of the **Assemble** tab; the **Place Constraint** dialog box is displayed.

5. Choose the **Mate** button from the dialog box and then, select the faces in the graphics window, as shown in Figure 9-22. As you select the second face, the Parallel key will move from the current location and will assemble with the existing assembly. Next, choose **Apply** button from the dialog box to apply the constraint.

6. Choose the **Mate** button and then select the edges, as shown in Figure 9-23. As soon as, you select the second edge to apply the constraint, the Parallel key will again move from its location and will be assembled with the rest assembly. Next, choose the **Apply** button to apply the constraint and then choose **Cancel** to exit the dialog box.

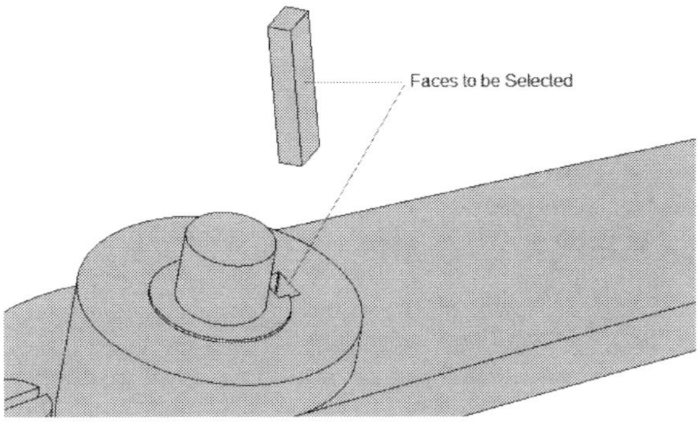

Figure 9-22 *Faces to be selected*

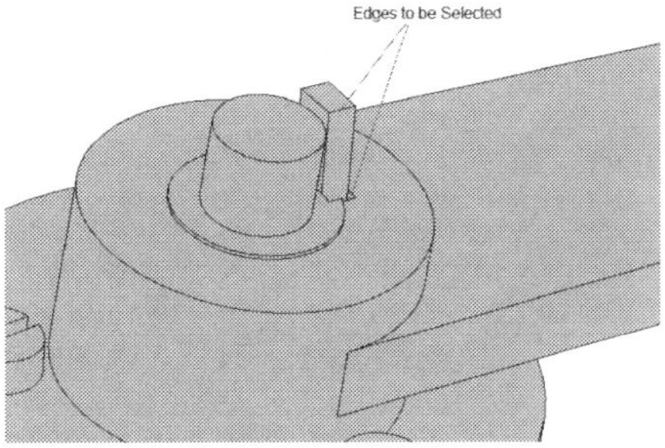

Figure 9-23 *Edges to be selected*

Assembling the Nut

Next, you need to assemble the Nut with the Shaft. Since the threaded portion of the Shaft has to be inserted inside the hole of the Nut, you need to use the **Insert** constraint to assemble these components.

1. Choose the **Place** tool from the **Component** panel to invoke the **Place Component** dialog box.

2. Double-click on the Nut; the Nut gets attached to the cursor.

3. Place the Nut at a location where it does not interfere with the existing components. Next, rotate the Nut using the **Free Rotate** tool such that the flat face of the Nut is visible in the current view.

4. Choose the **Constrain** tool from the **Relationship**s panel of the **Assemble** tab; the **Place Constraint** dialog box is displayed.

5. Choose the **Insert** button and then select the circular edge of the hole on the flat face of the Nut as the first face to apply the constraint.

6. Select the end face (not on the side of the chamfered edge) of the threaded feature of the Shaft. Choose the **Apply** button to apply the constraint and then choose **Cancel** to exit the dialog box. The final Butterfly Valve assembly is shown in Figure 9-24.

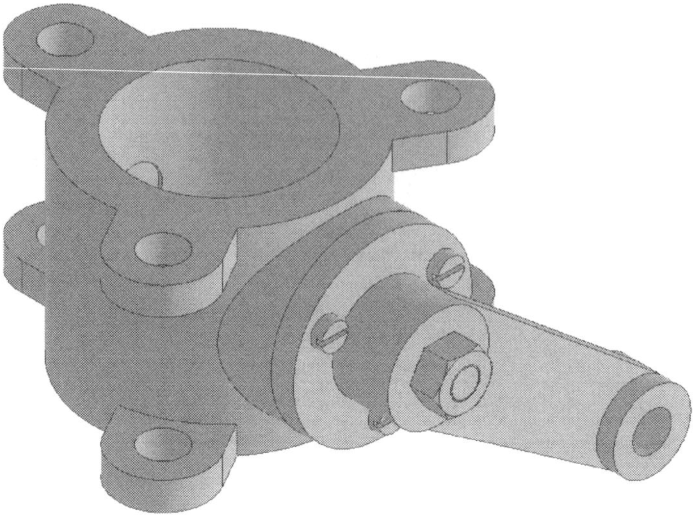

Figure 9-24 *Final Butterfly Valve assembly*

7. Save the assembly and close the file.

Tutorial 2

In this tutorial, you will create the components of a Plummer Block assembly. Note that you need to create all components as separate part files. After creating the components, place them in the assembly file and then assemble them. The views and dimensions of the components are shown in Figures 9-25 through 9-32. **(Expected time: 3 hr)**

Note

The orientation of the Casting that you will draw should match the orientation of the Casting shown in the assembly in Figure 9-25. This is because when you place the first component in the assembly file, it is placed on the same plane on which it was originally created in the part file. Since the Casting will be the first component to be placed in the assembly file, its base should be created on the XZ plane. The orientation of the other components also depends on the first component that you place in the assembly file.

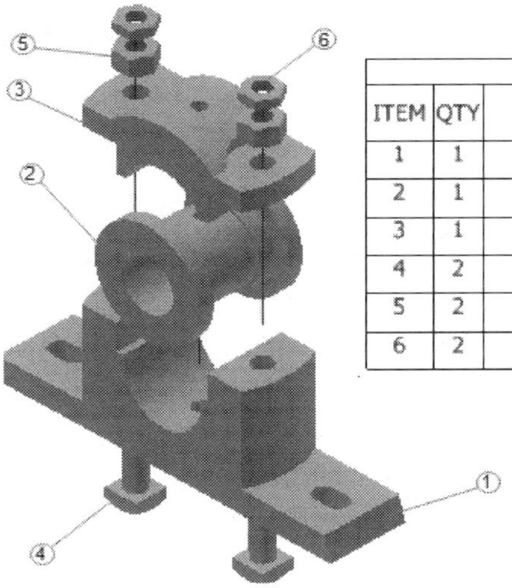

PARTS LIST		
ITEM	QTY	PART NUMBER
1	1	Casting
2	1	Brasses
3	1	Cap
4	2	Bolt
5	2	Nut
6	2	locknut

Figure 9-25 Exploded view of Plummer Block assembly

Figure 9-26 Half-sectioned isometric view of the Plummer Block assembly

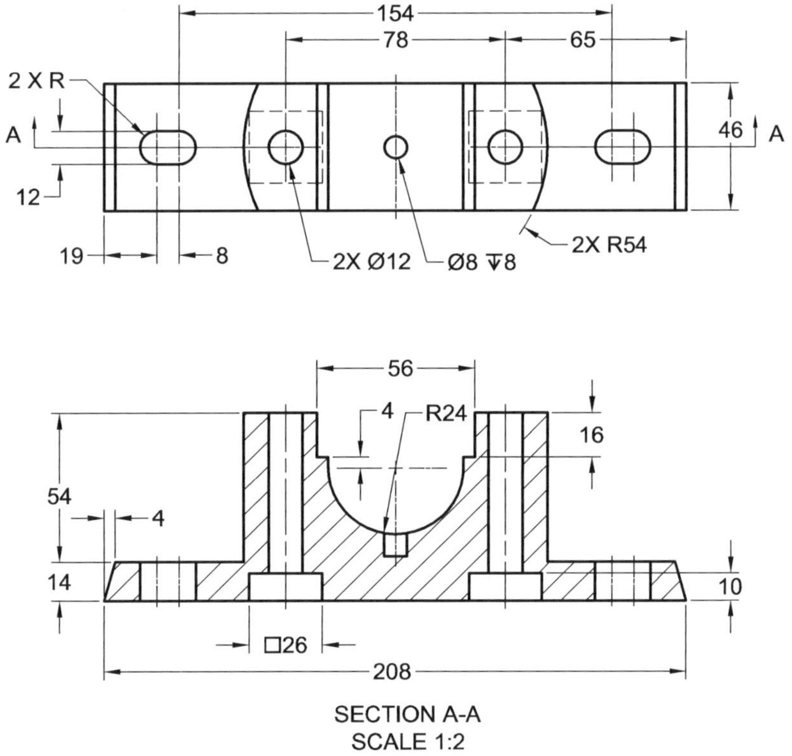

Figure 9-27 Views and dimensions of the Casting

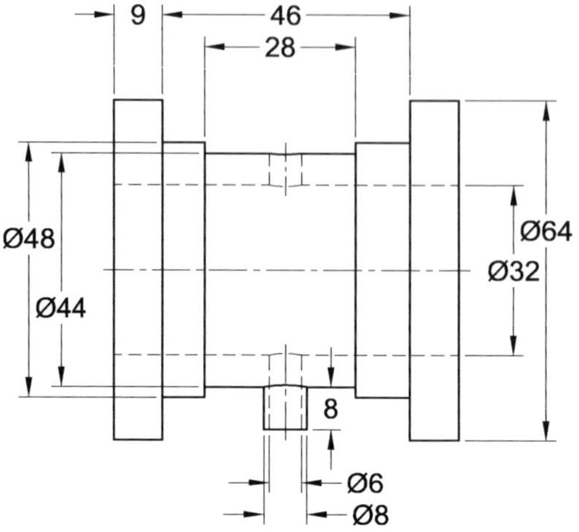

Figure 9-28 *Views and dimensions of the Brasses*

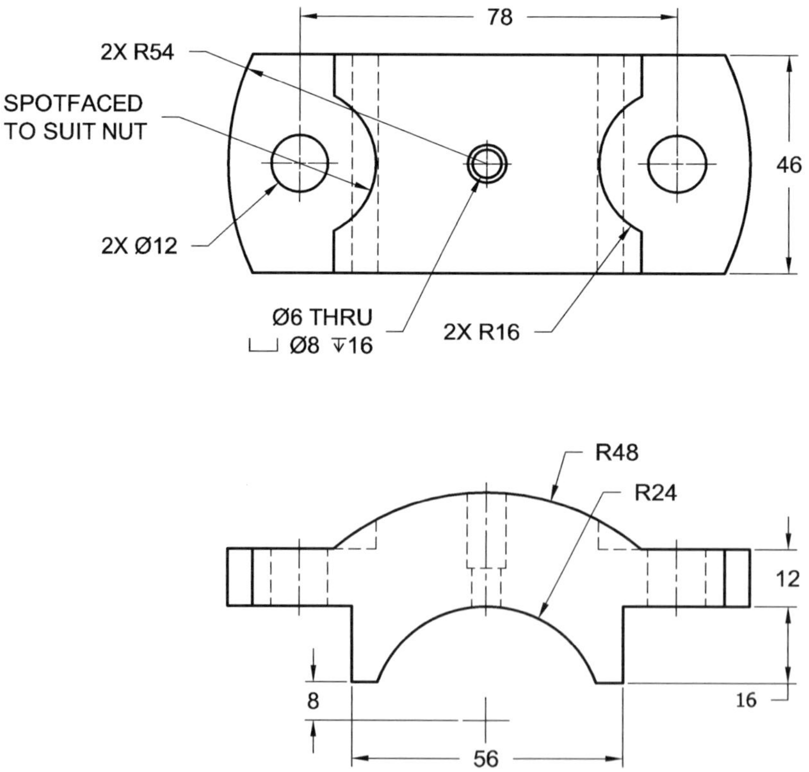

Figure 9-29 *Views and dimensions of the Cap*

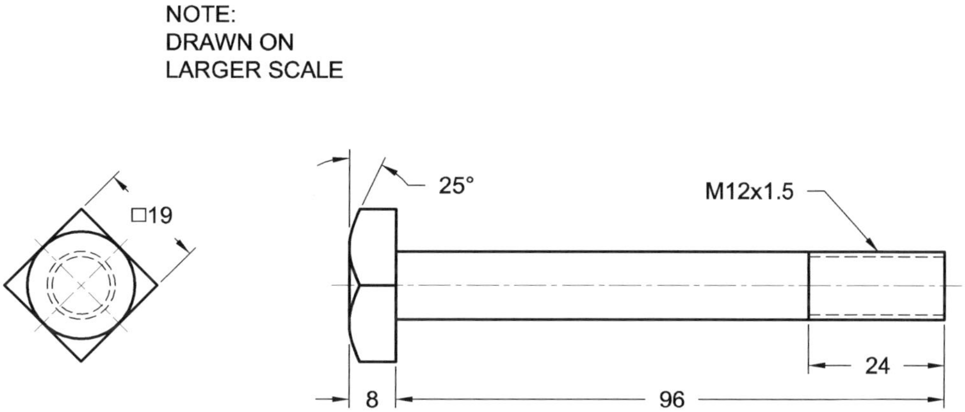

Figure 9-30 Views and dimensions of the Bolt

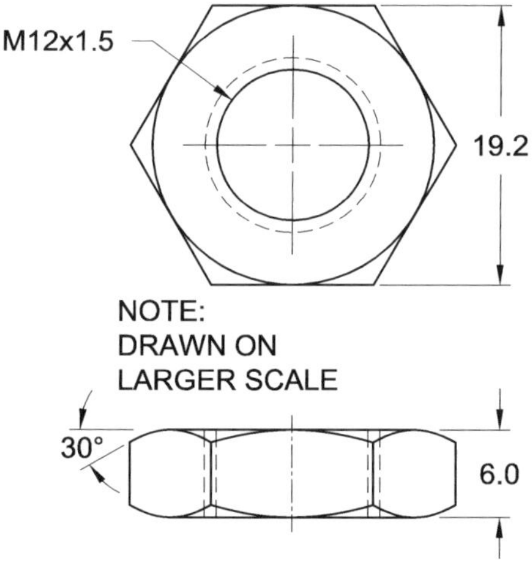

Figure 9-31 *Views and dimensions of the Lock Nut*

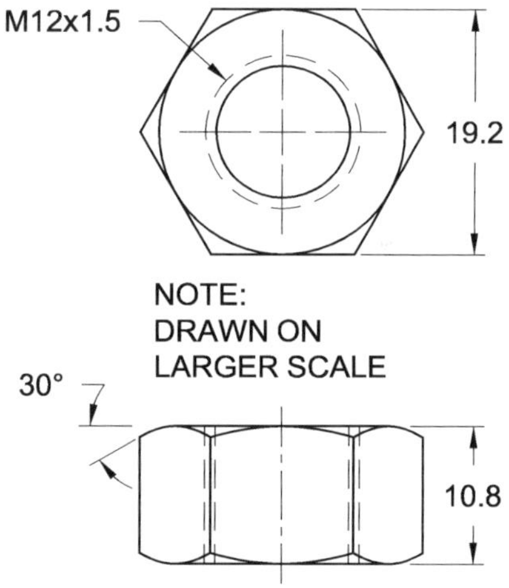

Figure 9-32 *Views and dimensions of the Nut*

The following steps are required to complete this tutorial:

a. Create all the components of the assembly as separate part files and save them in the *Plummer Block* folder at the location *Inventor_2020\c09*.
b. Start a new metric assembly file and then place the Casting and the Brasses by using the **Place** tool.
c. Assemble the two components using the assembly constraints.
d. Next, turn off the display of the Brasses and then place the Cap in the assembly. Assemble the Cap with the assembly.
e. Turn on the display of the Brasses.
f. Place two instances of the Bolt in the assembly file and then assemble them with the Casting and turn on the display of the Brasses.

g. Place two instances of the Nut and the Lock Nut. Assemble both the instances of the Nut with the Cap and then assemble the Lock Nut with the Nut.
h. Finally, turn on the display of the Brasses to complete the Plummer Block assembly.

Creating a New Project for the Assembly

1. Create a new folder with the name *Plummer Block* in the *c09* folder and set it as the current project folder using the procedure described in Tutorial 1.

Creating the Components

1. Create all components of the Plummer Block assembly as separate part files. Then, save the files with their respective names, refer to Figures 9-25 through 9-32. The files should be saved at the location *C:\Inventor_2020\c09\Plummer Block*.

Assembling the Casting and the Brasses

The Casting and the Brasses will be assembled using two **Mate** constraints. The **Mate** constraint is first applied between the axis of the snug on the Brasses and the axis of the hole available on the Casting. The **Mate** constraint is again applied between the cylindrical face of the Brasses and the cylindrical face of the Casting.

1. Start a new assembly file and save it with the name *Plummer Block* in the *Plummer Block* folder at the location *C:\Inventor_2020\c09*. The *Plummer Block* is the folder in which all the individual part files are saved.

2. Choose the **Place** tool from the **Component** panel of the **Assemble** tab to invoke the **Place Component** dialog box.

3. Double-click on the Casting; the Casting gets attached to the cursor and you are prompted to place the component. Next, you can use Marking menu to ground a component at the origin.

4. As you need to place only one instance of Casting, right-click in the graphics window; a Marking menu will be displayed. Choose **Place Grounded at Origin** from the Marking menu and again, right-click and choose **OK** from the Marking menu displayed. Next, place one instance of the Brasses in the current assembly file. Note that the Brasses should not be inserted as grounded component and its location should be such that it does not interfere with the Casting.

   While placing the component, if the orientation of the component is not same as shown in Figure 9-33 then rotate the component by 90 degrees to get the required orientation. To do so, right-click in the drawing area; a Marking menu will be displayed. Choose the required option from the Marking menu.

5. Choose the **Free Rotate** tool from the **Position** panel of the **Assemble** tab and rotate the Brasses such that its snug is visible in the current view.

6. Choose the **Constrain** tool from the **Relationships** panel of the **Assemble** tab; the **Place Constraint** dialog box is displayed.

7. In this dialog box, the **Mate** button is chosen by default. Select the axis of the snug as the first selection and axis of the hole on the Casting as the second selection, as shown in Figure 9-33.

8. Select the **Undirected** button from the **Solution** area.

9. Choose **Apply** from the **Place Constraint** dialog box to apply the constraint and exit the dialog box. Next, you need to apply another Mate constraint to make sure that the Brasses rest on the top of the cylindrical face of the Casting.

10. Move the Brasses up so that the bottom of the cylindrical face of the Brasses is visible. Next, invoke the **Place Constraint** dialog box. The **Mate** button is chosen by default in this dialog box.

11. Select the cylindrical face of the Brasses as the first selection and then select the cylindrical face of the Casting as the second selection, as shown in Figure 9-34. Next, choose the **OK** button in the **Place Constraint** dialog box; the constraints are applied, as shown in Figure 9-35.

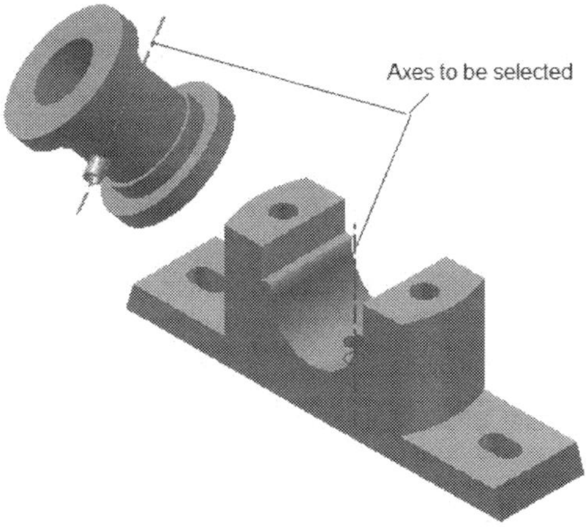

Figure 9-33 *Selecting the axes for applying the* ***Mate*** *constraint*

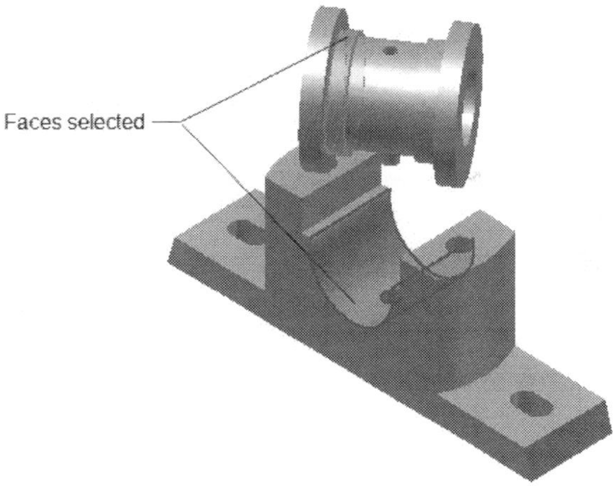

Figure 9-34 *Selecting the cylindrical faces of the Brasses and the Casting for applying the Mate constraint*

Assembling the Brasses and the Cap

After the Brasses are assembled, you now need to assemble the Cap with the assembly.

1. Place one instance of the Cap in such a manner that it does not interfere with the current assembly. Rotate the Cap in such a way that the inner cylindrical face of the Cap is visible.

2. Invoke the **Place Constraint** dialog box. In this dialog box, the **Mate** button is chosen by default. Select the axis of the hole on the cap as the first selection and then select the axis of the hole in the Brasses as the second selection, as shown in Figure 9-36. Next, choose the **Undirected** button from the **Solution** area.

3. Choose **OK** to apply the constraint and exit the **Place Constraint** dialog box.

Figure 9-35 *The Brasses after applying the* ***Mate*** *constraint*

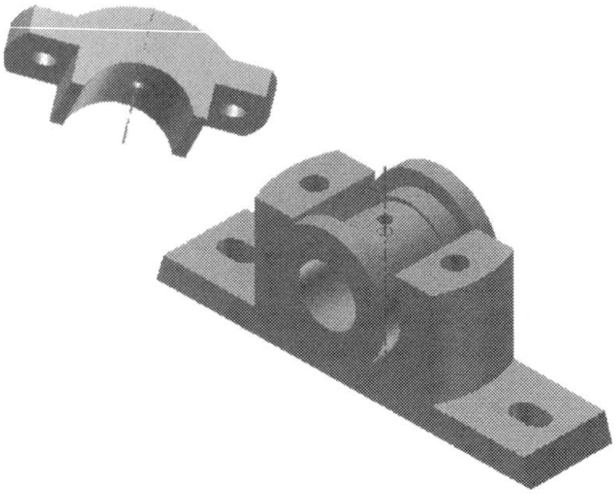

Figure 9-36 *Selecting the axes for applying the* ***Mate*** *constraint*

4. Now, you need to apply another Mate constraint between the cylindrical face of the Cap and the cylindrical face of the Brasses. To do so, move the Cap up in such a way that the cylindrical face of the Brasses is visible and then invoke the **Place Constraint** dialog box. In this dialog box, the **Mate** button is chosen by default. Move the cursor over the inner cylindrical face of the Cap and then right-click; a shortcut menu is displayed. Choose **Select Other**; the **Select Other** flyout is displayed. Click on the down arrow and then select the Face.

5. Similarly, select the top cylindrical face of the Brasses as the second selection, as shown in Figure 9-37. Next, choose **OK** from the **Place Constraint** dialog box to apply the constraint and close the dialog box. The assembly after the Cap is assembled is shown in Figure 9-38.

Figure 9-37 *Faces selected for applying the* ***Mate*** *constraint*

Figure 9-38 *The Cap after applying the* ***Mate*** *constraint*

Assembling the Bolts

There are two instances of the Bolts that have to be assembled in the current assembly. Since the Brasses are not required for assembling the Bolts or the Nuts, you can turn off their display. After turning off the display of the Brasses, you need to assemble the Bolts.

1. Turn off the display of the Brasses using the **Browser Bar**. Next, place two instances of the Bolt using the **Place** tool.

2. Change the display mode to Wireframe. Invoke the **Place Constraint** dialog box and then choose the **Insert** button from the **Type** area.

3. Select the circular edge on the top face of the base square feature of the Bolt as the first face to apply the constraint, refer to Figure 9-39.

4. Next, select the circular edge on the top face of the square cut of the bottom face of the Casting, see Figure 9-39. Choose the **Apply** button to apply the constraint.

5. Similarly, assemble the other Bolt and then change the display mode to shaded.

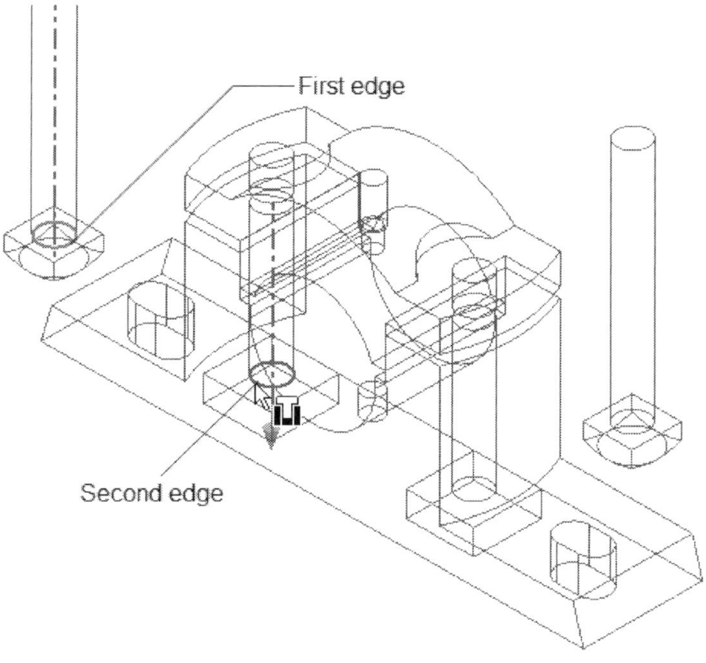

Figure 9-39 *Selecting the edges to apply the constraint*

6. You now need to apply the **Angle** constraint between the flat face of the head of the Bolt and the inner flat face of the slot so that the bolts do not rotate in the slots provided to them. To do so, invoke the **Place Constraint** dialog box and then choose the **Angle** button from the **Type** area of this dialog box.

7. Select the side planar face of the Bolt head and the inner planar face of the slot, as shown in Figure 9-40.

8. Choose the **Undirected Angle** button from the **Solution** area of the **Place Constraint** dialog box and enter **0** in the **Angle** edit box. Next, choose the **Apply** button in the **Place Constraint** dialog box.

9. Similarly, constrain the other Bolt in the assembly.

10. Turn on the visibility of the Brasses from the **Browser Bar** and change the display type to shaded.

11. Choose the **Home** button in the ViewCube. The home view of the assembly is enabled. The assembly after the Bolts are constrained is shown in Figure 9-41.

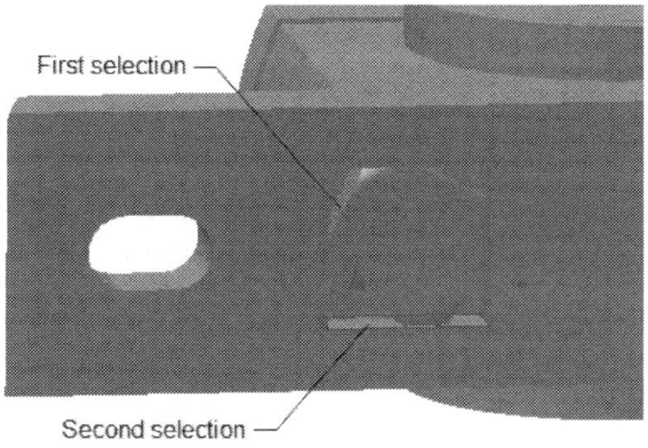

Figure 9-40 *Faces selected for applying the* ***Angle*** *constraint*

Figure 9-41 *The assembly after the Bolts are assembled*

Assembling the Nuts and the Lock Nuts

1. Place two instances each of the Nut and the Lock Nut using the **Place** tool.

2. Invoke the **Place Constraint** dialog box and choose the **Insert** button. Select the circular edge of the hole on the top face of one of the Nut as the first face to apply the constraint.

3. Select the circular edge of the left hole on the top face of the Cap as the second edge to apply the constraint. On doing so, the Nut is assembled with the Cap. Next, choose the **Apply** button to apply the constraint.

4. Select the circular edge of the hole on the top face of one of the Lock Nuts as the first face to apply the constraint.

5. Now, select the circular edge of the hole on the top face of the Nut that is assembled with the Cap to apply the constraint. Next, choose the **Apply** button to apply the constraint. The final Plummer Block assembly is shown in Figure 9-42.

6. Save the assembly by choosing the **Save** tool from the **Quick Access Toolbar**.

Figure 9-42 *Final Plummer Block assembly*

EXERCISES

Exercise 1

Create the components of the Drill Press Vice assembly and then assemble them, as shown in Figure 9-43. The dimensions of the components are shown in Figures 9-44 through 9-48. Create a folder with the name *Drill Press Vice* at the location *C:\Inventor_2020\c09* and save all the components and the assembly file in this folder. Assume the missing dimensions. You will use the bottom-up approach for creating this assembly.

(Expected time: 3 hr 15 min)

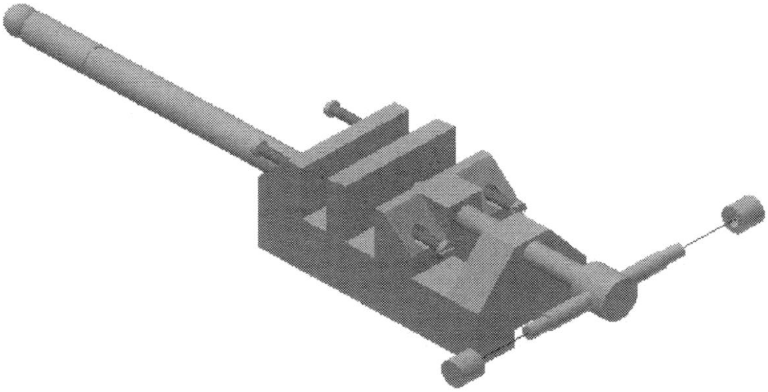

Figure 9-43 *Exploded view of the Drill Press Vice assembly*

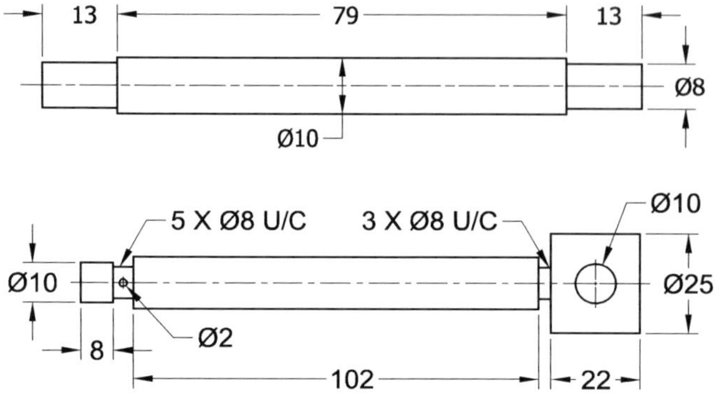

Figure 9-44 *Dimensions of the Clamp Screw Handle and Clamp Screw*

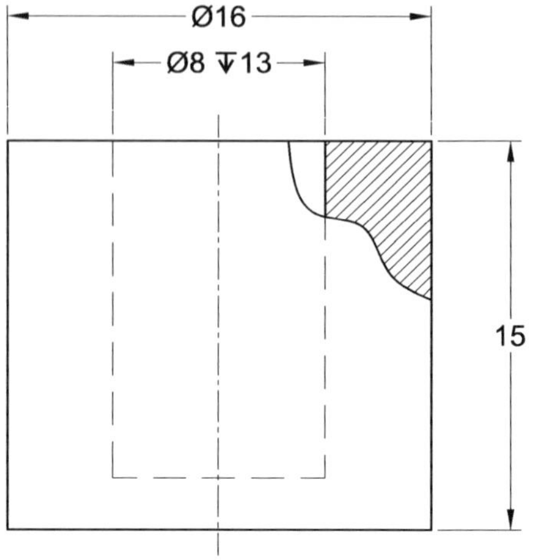

Figure 9-45 *Dimensions of the Handle Stop*

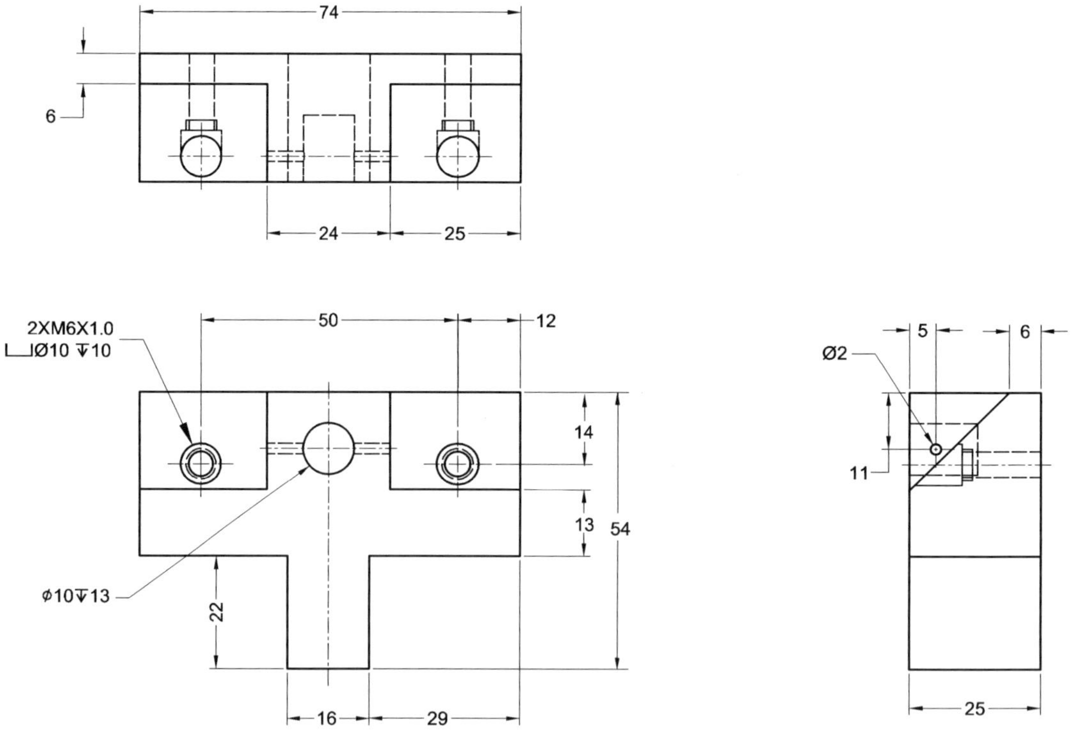

Figure 9-46 *Views and dimensions of the Movable Jaw*

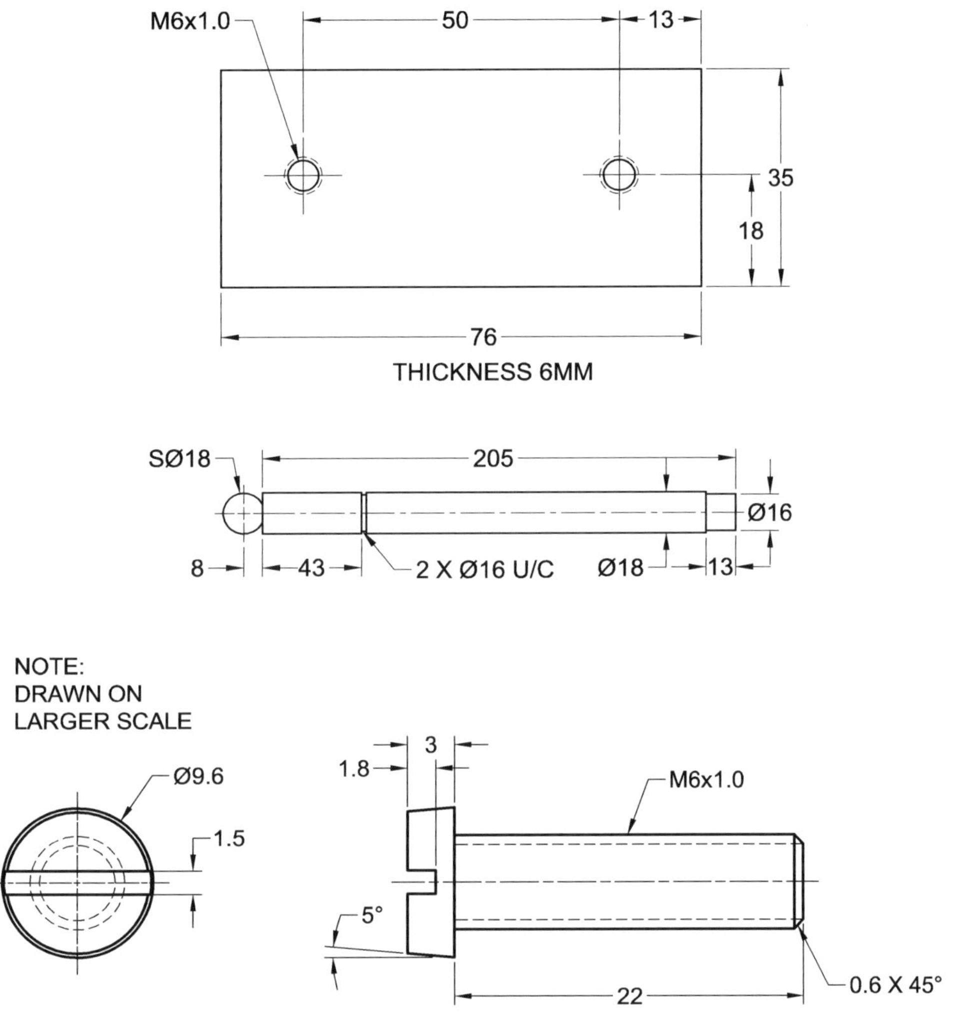

Figure 9-47 *Dimensions of the Jaw Face, Safety Handle, and Cap Screw*

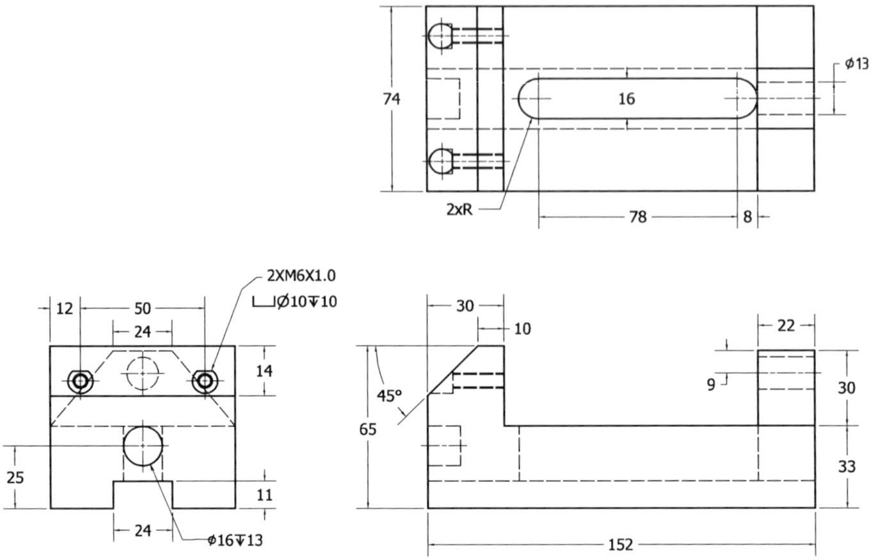

Figure 9-48 *Views and dimensions of the Base*

Chapter 10

Assembly Modeling-II

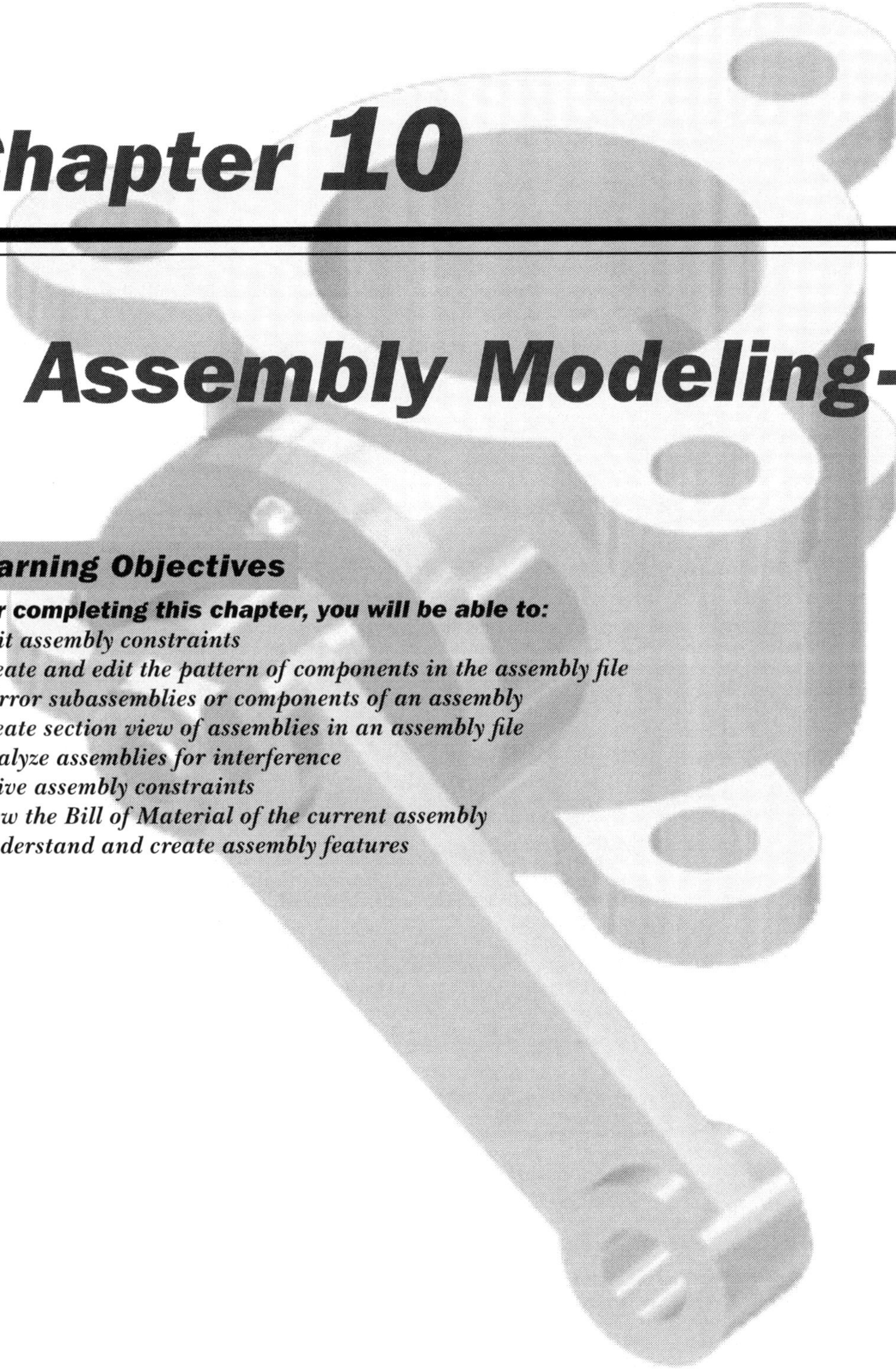

Learning Objectives

After completing this chapter, you will be able to:

- *Edit assembly constraints*
- *Create and edit the pattern of components in the assembly file*
- *Mirror subassemblies or components of an assembly*
- *Create section view of assemblies in an assembly file*
- *Analyze assemblies for interference*
- *Drive assembly constraints*
- *View the Bill of Material of the current assembly*
- *Understand and create assembly features*

EDITING ASSEMBLY CONSTRAINTS

Generally, after creating an assembly or during the process of assembling the components, you have to edit the assembly constraints that were used to assemble the components. The editing operations that can be performed on the assembly constraints include modifying the type of assembly constraint, the offset or angle values, the type of solution, or changing the component to which the constraint was applied.

EDITING COMPONENTS

Sometimes after assembling the components in an assembly, you need to edit the components. In Autodesk Inventor, you can edit the components by two methods. These methods are discussed next.

Editing Components in the Assembly File

The first method of editing components is to invoke the part modeling environment and the sketching environment in the assembly file and then edit the component. This method of editing components is similar to the top-down approach of assembly modeling.

Editing Components by Opening Their Part Files

The second method of editing the components is by opening their part files and making the necessary changes in them.

CREATING PATTERN OF COMPONENTS IN AN ASSEMBLY

Ribbon: Assemble > Pattern > Pattern

While creating assemblies, sometimes you have to assemble more than one instance of a component about a specified arrangement. For example, in case of a Butterfly Valve assembly, you have to assemble three instances of Screw with the Retainer and the Body (refer to Tutorial 1 of Chapter 9). All these three instances were recalled in the current assembly file and then assembled using the assembly constraint. Also, if you have to increase the number of holes in the Retainer and the Body from three to four, you will have to recall another instance of the Screw and insert it using the assembly constraint. However, this is a very tedious and time-consuming process. Therefore, to reduce the time for assembling the components, Autodesk Inventor has provided a tool for creating pattern of the components. You can use this tool to create circular or rectangular patterns.

MIRRORING SUBASSEMBLIES OR COMPONENTS OF AN ASSEMBLY

Ribbon: Assemble > Pattern > Mirror

Autodesk Inventor allows you to mirror assemblies or assembly components using the **Mirror** tool. You can use this tool to specify whether the mirrored components or subassemblies will be inserted in the current file or in a new assembly file.

CREATING ASSEMBLY SECTION VIEWS IN THE ASSEMBLY FILE

Sometimes, while assembling components in an assembly, some of the components are hidden behind the other components of the assembly. To visualize such components, Autodesk Inventor allows you to create the section views of the assembly. These section views are for reference only and the components are not actually chopped when you create the section views. You can create four types of section views: quarter section view, half section view, and three quarter section view by choosing the corresponding option from the **Section View** drop-down in the **Visibility** Panel of the **View** tab.

COPYING SUBASSEMBLIES OR COMPONENTS OF AN ASSEMBLY

Ribbon:	Assemble > Pattern > Copy

Similar to mirroring the components, you can also copy a subassembly or components of an assembly using the **Copy** tool.

DELETING COMPONENTS

You can delete the unwanted instances or the unwanted components from the assembly using the **Browser Bar**. In the **Browser Bar**, right-click on the unwanted component and choose **Delete** from the shortcut menu; the selected component will be deleted from the assembly.

EDITING THE PATTERN OF COMPONENTS

Autodesk Inventor allows you to edit the pattern of the components created using the **Pattern** tool. To edit the pattern of components, right-click on **Component Pattern** in the **Browser Bar** and choose **Edit** from the shortcut menu; the **Edit Component Pattern** dialog box will be displayed that can be used to edit the pattern.

DELETING ASSEMBLY CONSTRAINTS

You can delete the unwanted assembly constraints using the **Browser Bar**. To delete the assembly constraint, click on the > sign located on the left of the component in the **Browser Bar**. The **Origin** folder, along with all the constraints that are applied on the component, will be displayed. Right-click on the constraint to be deleted and choose **Delete** from the shortcut menu; the selected constraint will be deleted.

ANALYZING ASSEMBLIES FOR INTERFERENCE

Ribbon:	Inspect > Interference > Analyze Interference

Whenever you assemble the components of an assembly, no component should interfere with the other components of the assembly. If there is an interference between the components, it suggests that the dimensions of the components are incorrect or the components are not assembled properly. You will have to eliminate the interference in the assembly to increase the efficiency of the assembly and also eliminate the material loss.

SIMULATING THE MOTION OF COMPONENTS OF AN ASSEMBLY BY DRIVING ASSEMBLY CONSTRAINTS

Autodesk Inventor allows you to simulate the motion of the components of an assembly by driving the assembly joints and constraints. Remember that in the **Assembly** module, you can simulate the motion of the component using only one constraint at a time. However, you can create some relation parameters and equations for simulating the motion of the components using more than one constraint at a time.

CREATING POSITIONAL REPRESENTATIONS

Positional representations are the views of an assembly that represent assemblies in different component positions. For example, you can create a positional representation of an assembly in which the components are driven to a certain distance from their original assembly position. By default, every assembly has a main default positional representation. This positional representation represents the components at their default assembly position. You can create additional positional representations in which you can move the components from their default location by driving their constraints.

VIEWING THE BILL OF MATERIAL OF THE CURRENT ASSEMBLY

Ribbon: Manage > Manage > Bill of Materials

Autodesk Inventor allows you to view the Bill of Material of the current assembly in the assembly document itself. To view the Bill of Materials, choose the **Bill of Materials** tool from the **Manage** panel of the **Manage** tab; the **Bill of Materials** dialog box which lists the components of the current assembly in a tabular form will be displayed.

WORKING WITH ASSEMBLY FEATURES

Autodesk Inventor allows you to perform some metal cutting operations such as extrude, revolve, swept cuts, chamfer, and holes in an assembly file. Note that these operations are restricted only to the assembly file and are not performed on individual components. For example, if you create an extruded cut feature on a component in the assembly environment, the cut feature created in the assembly will not be created on the original component. As a result, this cut feature will be displayed only in the assembly environment and not in the original component file. Note that these operations are not restricted to a particular component, but extend to all the components of the assembly. For example, if you create a through-all cut feature in an assembly, the material will be removed not only from the component on which the sketch is created, but also from the components that come across the sketch.

TUTORIALS

Tutorial 1

In this tutorial, you will open the Butterfly Valve assembly created in Tutorial 1 of Chapter 9 and then analyze the assembly for interference. Next, you will delete the last two instances of the Screw and assemble the remaining instances by creating a pattern of the first instance.

(Expected time: 30 min)

The following steps are required to complete this tutorial:

a. Copy the *Butterfly Valve* folder from the *c09* folder to the *c10* folder.
b. Open the *Butterfly Valve.iam* file and analyze it for interference using the **Analyze Interference** tool.
c. Delete the two instances of the Screw assembled with the Retainer and then create two more instances using the **Pattern** tool.

Copying the Butterfly Valve Folder

In this tutorial, you will open the Butterfly Valve assembly created in *c09* folder. However, it is recommended that before opening the assembly file, you should copy the entire folder of the Butterfly Valve in the *c10* folder. This helps you keep the *c09* folder unaffected when you make changes in the Butterfly Valve Assembly. Therefore, first you will copy the *Butterfly Valve* folder in the *c10* folder and then open the *Butterfly Valve.iam* file from this folder.

1. Start a new session of Autodesk Inventor. Open the folder *c09* available at the location *C:\Inventor_2020*.

   You will notice that there is a folder with the name *Butterfly Valve* in *c09* folder. This is the folder where you have stored all part files and the assembly file of the Butterfly Valve.

2. Right-click on the *Butterfly Valve* folder and choose **Copy** from the shortcut menu.

3. Open the folder *C:\Inventor_2020\c10*. If this folder does not exist, you can create it using the **Create New Folder** button in the **Open** dialog box.

4. Right-click on the folder and paste the *Butterfly Valve* folder in it. Open the *Butterfly Valve* folder and then open the *Butterfly Valve.iam* file from it.

   The Butterfly Valve assembly is displayed on the screen.

Analyzing the Assembly for Interference

After opening the assembly, you will invoke the **Analyze Interference** tool and analyze the assembly for interference. There should be no interference in the assembly.

1. Choose the **Analyze Interference** tool from the **Interference** panel of the **Inspect** tab to display the **Interference Analysis** dialog box. In this dialog box, the **Define Set # 1** button is chosen by default. As a result, you are prompted to select the components to be added to the selection set.

2. Select Body from the graphics screen and then choose the **Define Set # 2** button from the dialog box. On doing so, you are again prompted to select the components to add to the selection set. Select the remaining components using the **Browser Bar**.

3. Choose the **OK** button; the **Analyzing Interference** dialog box is displayed. Also, you will notice that the system is analyzing the assembly for interference. After the analysis is complete, the **Autodesk Inventor Professional 2020** dialog box is displayed informing you that no interference is detected. Choose **OK** from this dialog box to exit.

Creating the Pattern of the Screw

While creating the Butterfly Valve assembly in Chapter 9, you assembled three instances of the Screw with the Retainer. You will retain the first instance of the Screw and delete the other two instances from the assembly. The other two instances will be assembled using the **Pattern** tool.

1. Select **Screw:2** from the **Browser Bar** and then press the SHIFT/CTRL key. Next, select **Screw:3** from the **Browser Bar**; you will notice that both the selected components are displayed in blue color in the **Browser Bar**. Also, the components are displayed with in a blue color on the graphics screen.

2. Press the DELETE key to delete two instances of the Screw.

   Since the holes on the Retainer are not visible in the current view, you need to turn off the visibility of the Arm.

3. Turn off the visibility of the Arm using the **Browser Bar**.

4. Choose the **Pattern** tool from the **Pattern** panel of the **Assemble** tab; the **Pattern Component** dialog box is invoked. In this dialog box, the **Component** button is chosen and you are prompted to select the component to be patterned.

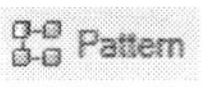

5. Select Screw as the component to be patterned. Choose the **Associated Feature Pattern** button from the **Feature Pattern Select** area of the **Associative** tab; you are prompted to select the feature pattern to associate to.

6. Select the hole on the lower part of the Retainer; two instances of the Screw are assembled with the two holes on the Retainer. Also, the display box on the right of the **Associated Feature Pattern** button displays **Circular Pattern1** which is the name of the pattern of holes on the Retainer.

Note

If you have created holes on the Retainer as circles while creating its basic sketch, you cannot use them to create associative component patterns because you can only associate the pattern to the feature pattern and not to the sketch pattern. In this case, you can create a non-associative pattern using the ***Circular*** *tab of the* ***Pattern Component*** *dialog box. However, as mentioned earlier, the pattern created using a tab other than the* ***Associative*** *tab will not be modified if the number of instances of the feature in the feature pattern are increased.*

7. Choose **OK** to create the pattern of the component and exit the **Pattern Component** dialog box.

   You will notice that the Screw is not displayed in the **Browser Bar** instead the **Component Pattern 1:1** node is displayed in it. If you click on the + sign on the left of this node, the three instances of the Screw with the name **Element:1**, **Element:2**, and **Element:3** are displayed in the **Browser Bar**.

8. Turn on the visibility of the Arm using the **Browser Bar**. Next, choose the **Save** tool from the **Quick Access Toolbar** to save the changes made in the assembly. The display of the **Browser Bar** after making all the changes in the assembly is shown in Figure 10-1.

9. Next, close the file.

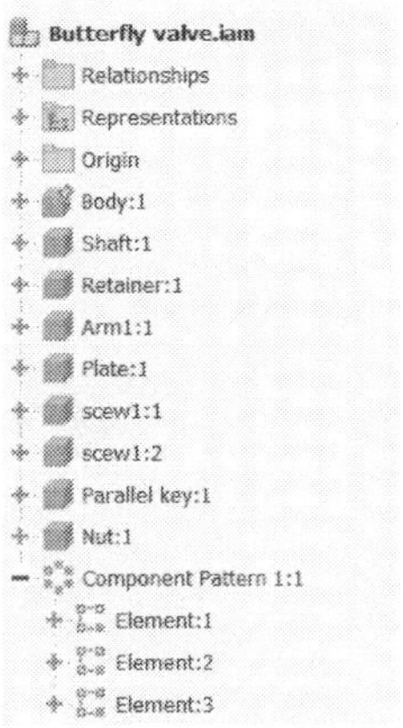

Figure 10-1 *Display of the* ***Browser Bar*** *for Tutorial 1*

Tutorial 2

In this tutorial, you will open the Drill Press Vice assembly created in *Exercise 1* of *c09* folder and then check the interference between the Base and the remaining components of the assembly. After checking the interference, you will drive the **Mate** constraint applied between the Clamp Screw and the Movable Jaw. **(Expected time: 30 min)**

The following steps are required to complete this tutorial:

a. Copy the Drill Press Vice assembly from the *c09* folder to the *c10* folder.
b. Open the *Drill Press Vice.iam* file and analyze it for interference.
c. Drive the **Mate** constraint applied between the vertical faces of the Jaw Face and the Base.

Copying the Drill Press Vice Assembly

As you do not want to modify the assembly created in *c09* folder, you need to copy the entire folder of the Drill Press Vice assembly to the *c10* folder. After copying the folder, you will open the assembly file and check components for interference.

1. Choose the **Open** tool from the **Launch** panel of the **Get Started** tab; the **Open** dialog box is displayed. Using this dialog box, open the folder *c09* from the location *C:\Inventor_2020*.

2. Right-click on the *Drill Press Vice* folder and then choose **Copy** from the shortcut menu.

3. Open the folder *C:\Inventor_2020\c10*. Right-click and choose **Paste** to paste the *Drill Press Vice* folder in the *c10* folder.

4. Open the *Drill Press Vice.iam* file from the *Drill Press Vice* folder.

Checking the Assembly for Interference

1. Choose the **Analyze Interference** tool from the **Interference** panel of the **Inspect** tab; the **Interference Analysis** dialog box is invoked.

   In this dialog box, the **Define Set # 1** button is chosen by default. As a result, you are prompted to select the components to add to the selection set.

2. Select Base from the graphics screen. Next, choose the **Define Set # 2** button from the dialog box and then select the remaining components in the **Browser Bar**.

3. Choose **OK** from the **Interference Analysis** dialog box; the **Analyzing Interference** dialog box is displayed informing you that the interference is being analyzed.

Driving the Constraint to Simulate the Motion of the Assembly

The Clamp Screw Handle and the two instances of the Handle Stop were assembled with the Clamp Screw using assembly constraints. Therefore, when you simulate the Clamp Screw by driving its constraint, you will notice that the Clamp Screw Handle and both the instances of the Handle Stop will also move along with the Clamp Screw.

1. Click on the + sign located on the left of the **Jaw Face** node in the **Browser Bar**; the node is expanded and the **Origin** folder along with various constraints applied to it is displayed.

2. Move the cursor over the Mate constraint; the mating faces of the Jaw Face:1 and the Jaw Face:2 are highlighted on the graphics screen. This is done to ensure that the constraint you selected is the correct one. Next, right-click on the Mate constraint, and then choose **Drive** from the shortcut menu; the **Drive** dialog box is displayed.

3. Enter **10** and **60** in the **Start** and **End** edit boxes respectively as the start and end values of the simulation.

4. Choose the **More (>>)** button to expand the dialog box. Select the **Start/End/Start** radio button from the **Repetitions** area and then enter **2** in the edit box provided in the same area.
   As you enter **2** in the edit box, two cycles of simulation of the assembly are created. The first cycle will be from the start position to the end position and the second cycle will be from the end position to the start position.
5. Choose the **Forward** button; you will notice that there is horizontal simulation of the Jaw face. Also, other components assembled to it will move along with it. As there are two repetitions, first the components will move 30 mm away from the Movable Jaw and then move back to the start position.

6. Exit the **Drive** dialog box by choosing the **Cancel** button. Save the changes made to the assembly and then close the file.

Tutorial 3

In this tutorial, you will create the components of the Double Bearing assembly and then assemble them, as shown in Figure 10-2. Figure 10-3 shows the required positional representation of the assembly. Use the **Pattern** tool while assembling the Bolts. The dimensions of various components are given in Figures 10-4 through 10-6. After assembling the components, drive the **Insert** constraint of the first Bolt such that the remaining three instances are also simulated. Create a positional representation of the assembly with the Bolts at the new location. **(Expected time: 2 hrs)**

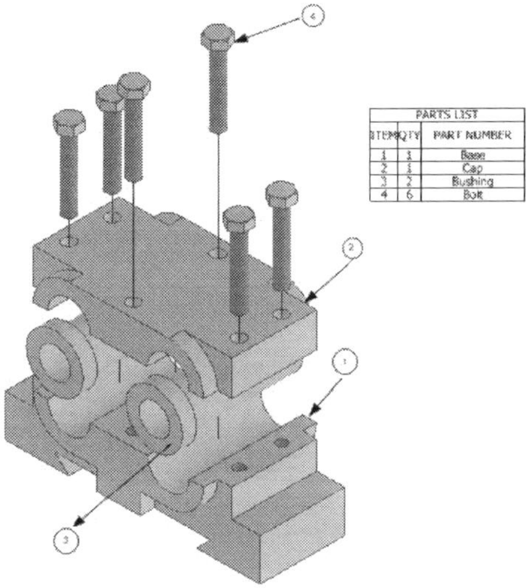

Figure 10-2 *Exploded view of the Double Bearing assembly*

Figure 10-3 *Required positional representation of the Double Bearing assembly*

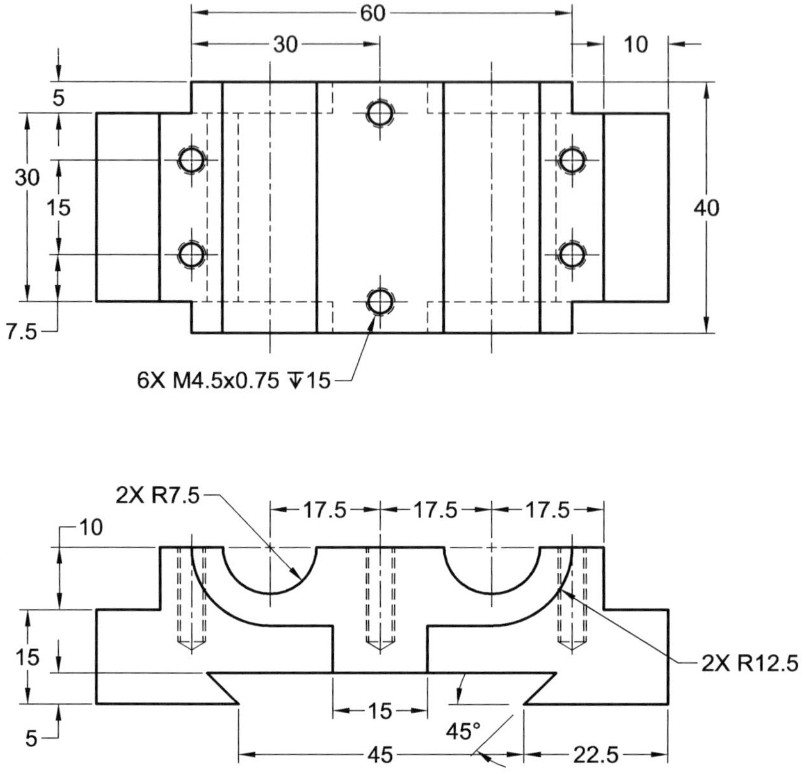

Figure 10-4 *Views and dimensions of the Base*

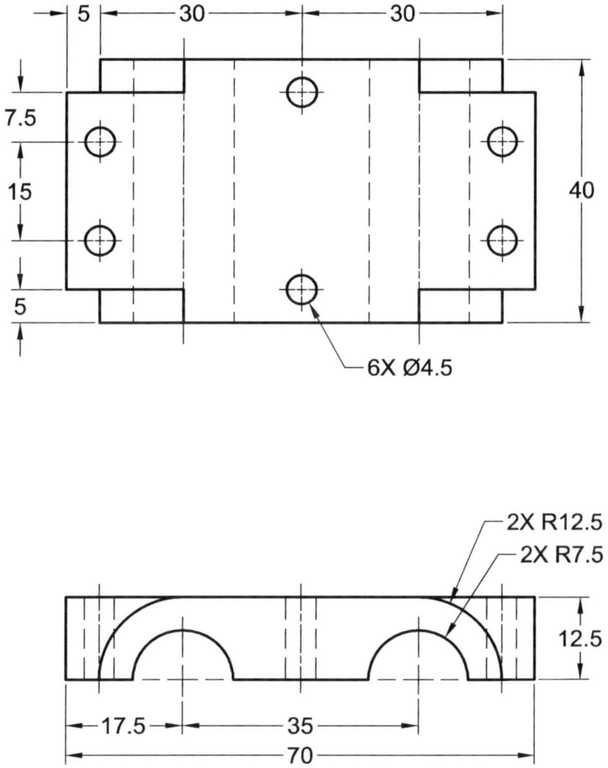

Figure 10-5 *Views and dimensions of the Cap*

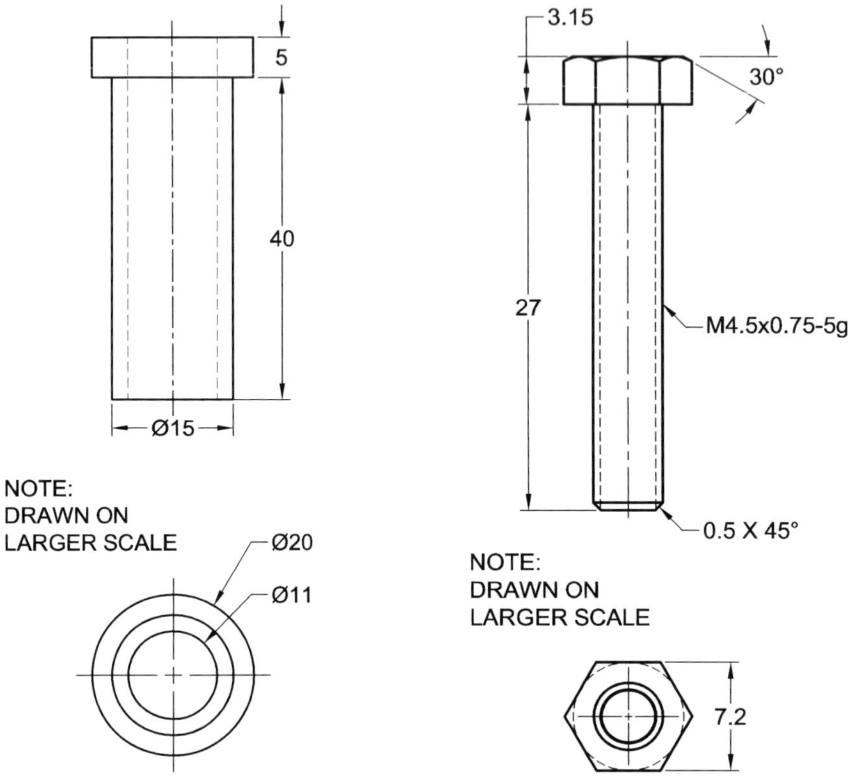

Figure 10-6 *Dimensions of the Bushing and Bolt*

The following steps are required to complete this tutorial:

a. Create a folder with the name *Double Bearing* inside the *c10* folder. Create all components of the Double Bearing assembly and save them in this folder.
b. Open a new assembly file and assemble the components of the Double Bearing assembly. Only two instances of Bolt should be assembled and the rest should be assembled by patterning.
c. Create a new positional representation of the assembly.
d. Drive the **Insert** constraint applied between one of the Bolts and the Cap so that the Bolts are moved to a new location in the current positional representation.

Creating Components

1. Create a folder with the name *Double Bearing* at the location *C:\Inventor_2020\c10* and then create all components in individual part files and save them in this folder.

2. Open a new assembly file and save it with the name *Double Bearing.iam* at the location *C:\Inventor_2020\c10\Double Bearing*.

Assembling Components

The first component that has to be restored is the Base. Next, you need to restore the Cap. Then, you need to assemble the base and the cap using the assembly constraints. Next, you need to assemble two instances of the Bushing and then two instances of the Bolt. The remaining instances of the Bolt will be assembled using the **Pattern** tool.

1. Place one instance each of the Base and the Cap in the assembly file by using the **Place** tool and make the base grounded. If required, you can choose the **Free Rotate** tool from the Marking Menu to retain the orientation of the Assembly, shown in Figure 10-7.

2. Assemble these components using the **Constrain** tool. The assembly after assembling the Base and the Cap is shown in Figure 10-7.

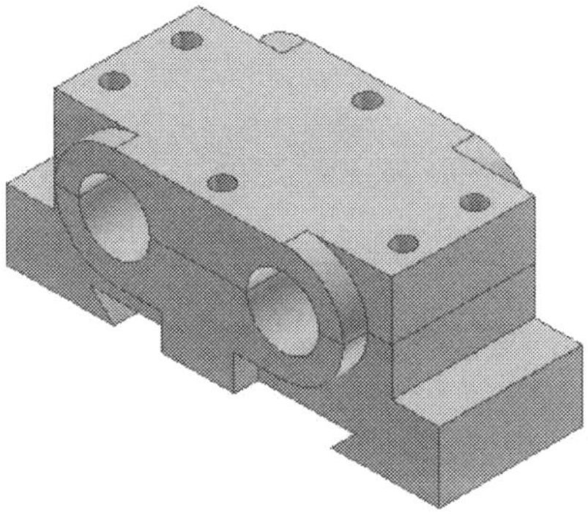

Figure 10-7 *Assembly after assembling the Base and the Cap to it*

3. Place two instances of the Bushing and then assemble them using the **Constrain** tool, refer to Figure 10-8.

4. Similarly, place two instances of the Bolt and then assemble them using the **Constrain** tool, as shown in Figure 10-8. You also need to make sure that you place the bolts in the parent holes in the cap part of the model. To check the parent or the primary holes of the cap, check the part file.

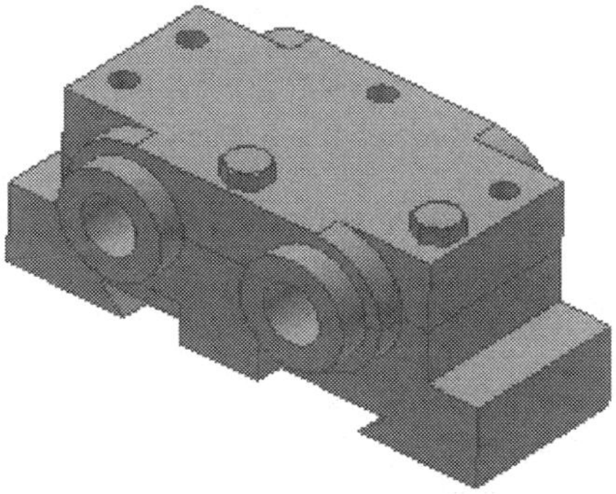

Figure 10-8 *Assembly after two instances of Bolts are assembled with the Base and the Cap*

It is presumed that one of the four holes at the corners of the Cap was created and the other three were patterned. Similarly, one of the holes in the middle of the Cap was created and the other was patterned. Since you have not created all the six holes using the single pattern, you need to use the **Pattern** tool twice. First time, the tool will assemble the Bolt on the three holes at the corners and the second time, it will assemble the Bolt on the remaining hole in the middle of the Cap.

5. Choose the **Pattern** tool from the **Pattern** panel of the **Assemble** tab; the **Pattern Component** dialog box is invoked. Also, you are prompted to select the component to be patterned.

6. Select the Bolt at the upper left corner of the Cap.

7. Choose the **Associated Feature Pattern** button from the **Feature Pattern Select** area; you are prompted to select the feature pattern to associate to.

8. Select one of the three holes at the corners of the Cap; the three instances of the Bolt are assembled at three holes.

Note

The pattern of bolts created depends on the pattern of hole created on the cap and the location of the first hole on the cap. Therefore, you need to be careful while specifying the location of the first instance of the hole on the cap in the ***Part*** *environment.*

9. Choose **OK** to assemble the remaining three instances of the Bolt and exit this dialog box.

10. Invoke the **Pattern Component** dialog box again and select the Bolt assembled with the hole in the middle of the Cap.

11. Choose the **Associated Feature Pattern** button from the **Feature Pattern Select** area; you are prompted to select the feature pattern to associate to.

12. Select the other hole in the middle of the Cap and choose **OK**. The final Double Bearing assembly is shown in Figure 10-9.

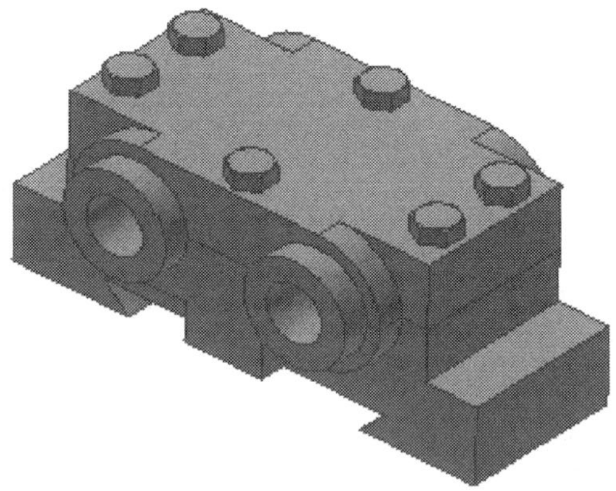

Figure 10-9 *Double Bearing assembly*

13. Choose the **Save** tool from the **Quick Access Toolbar** to save the assembly.

Creating the Positional Representation of the Assembly

As mentioned in the tutorial description, you need to create the positional representation of the assembly with the Bolts moved to an offset position of 30 mm. To do so, you first need to create a positional representation and then drive the constraints of the Bolts such that the

Bolts are moved to a new location in the current positional representation. The positional representations are created using the **Browser Bar**.

1. Click on the + sign located on the left of the **Representations** node in the **Browser Bar** to expand the tree view.

2. Right-click on **Position** in the **Browser Bar** and choose **New** from the shortcut menu; the **Position** changes to **Position : Position 1** node and a + sign is added on its left.

3. Click on the + sign located on the left of **Position : Position1** in the **Browser Bar**; the tree view expands. You will notice that a check mark is displayed on the left of **Position1**, suggesting that this is the current representation.

Driving the Constraint of the Bolt

When you drive the constraint of the first Bolt at the lower right corner of the cap, you will notice that the remaining three instances at the corners of the cap also simulate along with the first Bolt because the remaining three instances were assembled using the **Pattern** tool. This tool will force the three instances to behave in the same way as the original Bolt does.

1. Click on the + sign located on the left of **Component Pattern 1** in the **Browser Bar**. You will notice that four instances of the Bolts are displayed with the name **Element:1**, **Element:2**, **Element:3**, and **Element:4**.

2. Click on the + sign on the left of **Element:1**; **Bolt:1** is displayed. Similarly, click on the + sign on the left of **Bolt:1** to display the **Insert** constraint.

3. Right-click on the **Insert** constraint, and then choose **Drive** from the shortcut menu; the **Drive** dialog box is displayed.

4. Enter **30** in the **End** edit box and then choose the **More** button to expand the dialog box.

5. Select the **Start/End/Start** radio button from the **Repetitions** area and then enter **2** in the edit box available in this area.

6. Choose the **Forward** button; all four bolts at the corners of cap will be simulated and moved to a distance of 30 mm in the upward direction. All the four bolts are then moved back to their original positions without any pause between the cycles.

   As you need to create a positional representation of the assembly with the bolts at an offset of 30 mm from the original location, you need to stop the movement of the Bolts at the top most position. To do this, you need to modify the value in the edit box in the **Repetitions** area of the **Drive Constraint** dialog box.

7. Enter **1** in the edit box of the **Repetitions** area and then choose the **Forward** button; the Bolts move up to a distance of 30 mm in the upward direction. Figure 10-10 shows the assembly with four bolts at the new position.

8. Choose the **OK** button to exit the **Drive** dialog box. Choose **Yes**, if the **Autodesk Inventor Professional** message box is displayed. This message box informs you that the value of the constraint must be overridden in the current positional representation to preserve it.

9. Choose the **Save** tool from the **Quick Access Toolbar**; the **Autodesk Inventor Professional 2020** dialog box is displayed and you are informed that you cannot save the assembly when the assembly is in the positional representation.

Figure 10-10 *Position of the bolts after using the **Drive** option*

10. Choose **OK** from this dialog box to restore the master representation and save the assembly file.

11. Close the assembly document.

EXERCISE

Exercise 1

Open the Plummer Block assembly created in Tutorial 2 of Chapter 9 and then create a design view representation with the name Plummer Block, refer to Figure 10-11. After creating the design view, analyze the assembly for interference and then simulate the motion of the two Bolts. Note that the bolts should move in a downward direction. **(Expected time: 30 min)**

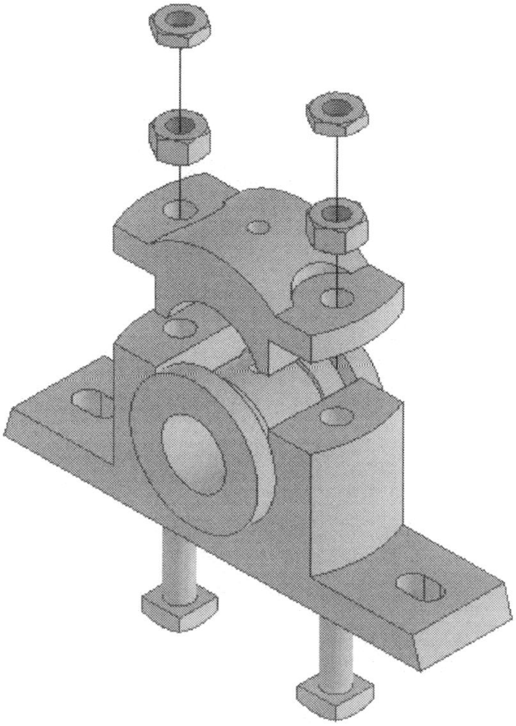

Figure 10-11 *Design view representation of the Plummer Block assembly*

This page is intentionally left blank

Chapter 11

Working with Drawing Views-I

Learning Objectives

After completing this chapter, you will be able to:
- *Understand the use of drawing module*
- *Understand various types of drawing views in Autodesk Inventor*
- *Assign different hatch patterns to different components in assembly section views*
- *Exclude components in assembly section views*

THE DRAWING MODULE

After creating a solid model or an assembly, you need to generate their drawing views. Drawing views are the two-dimensional (2D) representations of a solid model or an assembly. Autodesk Inventor provides you with a specialized environment for generating drawing views. This specialized environment is called the **Drawing** module and has only those tools that are related to drawing views. As mentioned earlier, all modules of Autodesk Inventor are bidirectionally associative. This property ensures that changes made in a part or an assembly are reflected in drawing views. Also, changes in the dimensions of a component or an assembly in the **Drawing** module are reflected in the part or assembly file.

TYPES OF VIEWS

In Autodesk Inventor, you can generate various types of views from a model, assembly, or presentation. Additionally, you can also draft a view using the sketcher entities. The technique of generating drawing views from models, assemblies, and presentations is called generative drafting. This is because you generate the drawing views. The technique of drafting a drawing view using the sketcher entities is called interactive drafting. The types of drawing views that you can generate are discussed next.

Base View

The base view is the first view generated in the drawing sheet. This view is generated using the original model, assembly, or presentation. The base view is an independent view and is not affected by changes in any other view in the drawing sheet. Most of the other views in the sheet will be generated taking this view as the parent view.

Projected View

The projected view is generated taking any of the existing views as the parent view. This view is generated by projecting the lines normal to the parent view or at an angle to the parent view to generate a 3D view. If the lines are projected normal to the parent view, the resulting view will be an orthographic view such as top view, front view, side view, and so on. If the lines are projected at an angle, the resulting view will be a 3D view such as an isometric view. In this view, you can visualize the X, Y, and Z axes of the model. These views are 2D representations of a three-dimensional (3D) model.

Auxiliary View

An auxiliary view is a drawing view that is generated by projecting the lines normal to a specified edge of an existing view.

Section View

A section view is generated by chopping a part of an existing view using a plane and then viewing the parent view from a direction normal to the section plane.

Detail View

A detail view is used to display the details of a portion of an existing view. You can select the portion whose detail view has to be shown in the parent view. The portion that you have selected will be magnified and placed as a separate view. You can control the magnification of the detail view.

Overlay View
An overlay view is used to display an alternate position of the components in an assembly. It uses positional representations created in the assembly environment for generating the drawing view.

Broken View
A broken view is used to display a component by removing a portion of it from the middle and keeping the ends of the drawing view intact. This type of view is used for displaying the components whose length to width ratio is very high. This means that either the length is more as compared to the width or the width is more as compared to the length. The broken view will break the view along the horizontal or vertical direction such that the drawing view fits the area you require. Note that in these views, the dimension of the edge that is broken will still be displayed as the actual value. However, this dimension will have a broken symbol suggesting that the dimension value is for the edge that is broken in the view.

Break Out View
A break out view is used to remove a part of the existing view and to display the area of the model or the assembly behind the removed portion. This type of view is generated using a closed sketch that is associated with the parent view.

Slice
A slice view is used to indicate important portions of a part or an assembly file as a zero depth section. It is generated on a target view by creating a sketch for the material to be removed on the source view.

Crop
A crop view is used to crop an existing view enclosed in a closed sketch associated to that view. The portion of the view that lies inside the associated sketch will be retained and the remaining portion will be removed. You can also crop a view by creating the rectangular trap by using the **Crop** tool. In this method the portion that lies inside the rectangular trap will be retained.

ASSIGNING DIFFERENT HATCH PATTERNS TO COMPONENTS IN ASSEMBLY SECTION VIEWS
Whenever you generate the section views of an assembly, by default, similar hatch patterns are assigned to all of them. Although the angle of hatching lines between the adjacent components is different yet it creates confusion if the assembly has a number of components.

EXCLUDING COMPONENTS FROM ASSEMBLY SECTION VIEWS
To prevent the components from being sectioned, click on the + sign located on the left of the section view; the name of the assembly will be displayed in the **Browser Bar**. Click on the + sign located on the left of the assembly name to display all the components of the assembly in the **Browser Bar**. Now, hold the CTRL key and then use the left mouse button to select all the components that you want to exclude from sectioning. Once all the components are selected, they will be displayed in a blue background in the **Browser Bar**. Right-click on any of the selected

components to display the shortcut menu. Choose **Section Participation > None** from the shortcut menu; all the selected components will be excluded from the section view.

TUTORIALS

Tutorial 1

In this tutorial, you will generate the top view, full sectioned front view, and isometric view of the sectioned front view of the model created in Tutorial 2 of the c07. Use the JIS standard template file for generating the views. **(Expected time: 30 min)**

The following steps are required to complete this tutorial:

a. Copy the model of Tutorial 2 of the *c07* folder to the current folder.
b. Open a JIS template file and generate the base view using the **Base** tool.
c. Generate the section view by sketching the section plane.
d. Use the **Projected** tool to project lines at an angle from the section view to generate the isometric view.

Copying the Model to the Current Folder

Before generating the drawing view of the model, it is important to copy the model to the current folder. This is necessary because when you open the drawing file next time, the component will be searched in the current *c11* folder. If the component is not available in the current folder, the **Resolve Link** dialog box will be displayed. This dialog box will prompt you to specify the location and path of the component file. Therefore, all the components or the assemblies should be copied into the current folder, or the drawing file should be saved in the folder in which the component and assembly file are located.

1. Create a folder with the name *c11* at the location *C:\Inventor_2020* and then copy the *Tutorial 2.ipt* file from the location *C:\Inventor_2020\c07* to the *c11* folder. Next, rename this file as *Tutorial 1.ipt*.

Starting a New Drawing File

As mentioned in the tutorial description, you need to use the JIS standard template for generating the drawing views. Therefore, you will use the *JIS.idw* file for generating the drawing views.

1. Start a new session of Autodesk Inventor and choose the **New** tool from the **Launch** panel of the **Get Started** tab to invoke the **Create New File** dialog box.

2. Choose the **Metric** tab and then double-click on the **JIS.idw** option to open a JIS standard drawing file, see Figure 11-1.

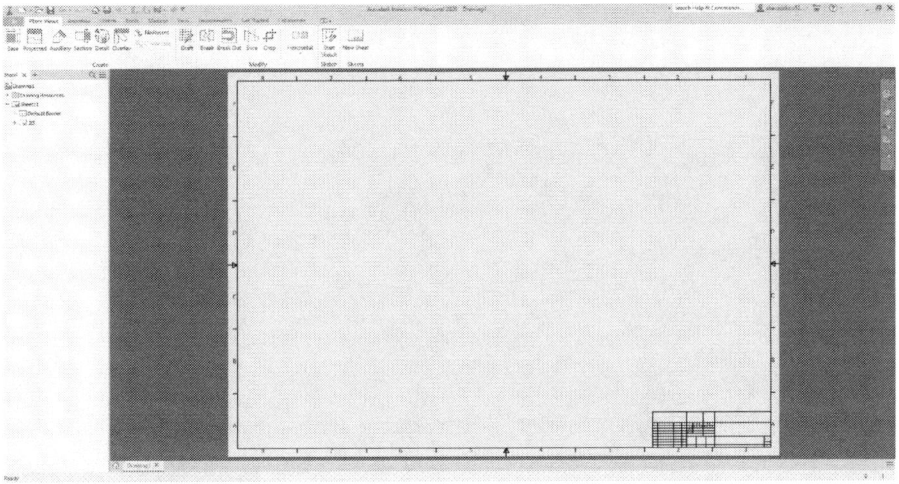

Figure 11-1 *Screen display of the JIS standard drawing file*

Generating the Base View

As mentioned earlier, the base view is the first view in the drawing sheet. Once you have generated the base view, you can use it as the parent view for generating other views. The base view is generated using the **Base View** tool.

1. Choose the **Base View** tool from the Marking menu which is displayed when you right-click anywhere in the graphics window; the **Drawing View** dialog box is displayed.

   Preview of the drawing view is not displayed on the sheet because you have not selected any part file. Therefore, first you need to select the part file for which the drawing view will be generated.

2. Choose the **Open an existing file** button on the right of the **File** drop-down list in the **Component** tab of the **Drawing View** dialog box; the **Open** dialog box is displayed.

3. In this dialog box, select *Tutorial1.ipt* from *C:\Inventor_2020\c11* and then choose the **Open** button.

   You will notice that the preview of the drawing view with default orientation is displayed. Also, its projected view gets attached to the cursor. The projected view moves as you move the cursor and will be generated at the point that you specify in the graphics window.

4. Modify the scale value to **1.5 : 1** in the **Scale** edit box. Make sure that only the **Hidden Line** button is chosen from the **Style** area in the **Component** tab of the **Drawing View** dialog box.

5. Next, choose the **OK** button to exit the drawing view dialog box; the drawing view is created.

6. Select and drag the drawing view to the top left corner of the sheet, see Figure 11-2.

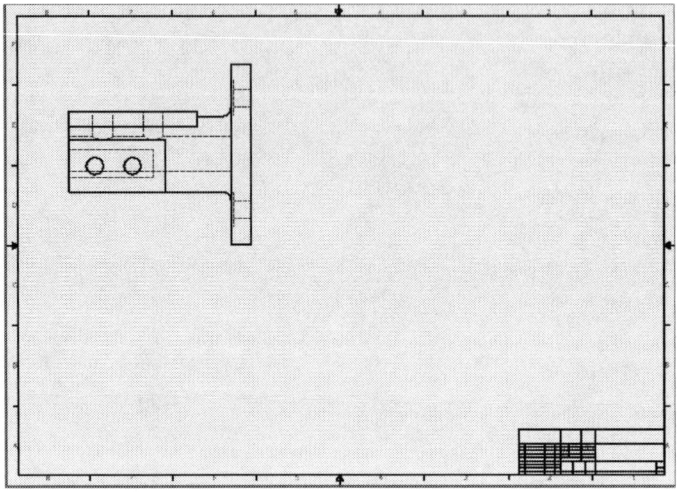

Figure 11-2 *Drawing sheet with the drawing view*

Note

The ***Drawing View*** *dialog box if hinders specifying point on the sheet can be moved by dragging it.*

Generating the Section View

The section view can be generated using the **Section** tool. To generate the view using this tool, first you need to select the drawing view that has to be sectioned and then define the section plane. But, if you use the shortcut menu that is displayed by right-clicking on the base view in the **Browser Bar** or in the drawing sheet, you do not need to select the drawing view as it is already selected. Therefore, you will use this shortcut menu to generate the section view.

1. Move the cursor over the base view on the sheet; a the red dotted bounding box is displayed. Right-click and choose **Create View > Section View** from the shortcut menu; the cursor changes into a sketch cursor and you are prompted to enter the endpoints of the section line.

2. Move the cursor close to the midpoint of the extreme left vertical edge of the base view; the cursor snaps at the midpoint and turns green, refer to Figure 11-3.

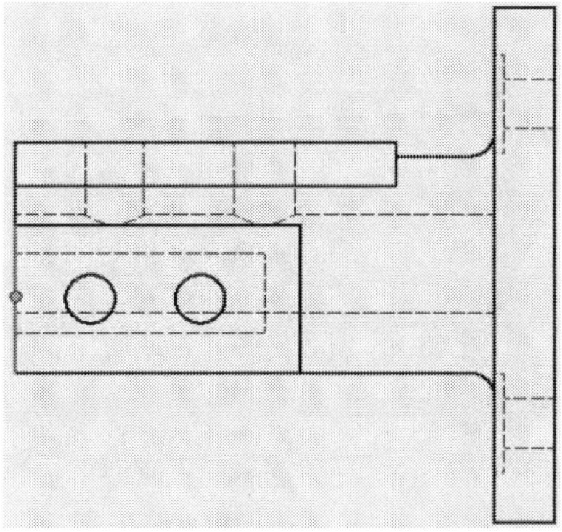

Figure 11-3 *Cursor snapping at the midpoint of the left vertical edge*

3. Move the cursor horizontally toward the left of the view. You will notice that an imaginary horizontal line is drawn from the midpoint of the left vertical edge. This is due to the temporary tracking option.

4. Click to specify a point after slightly moving the cursor horizontally toward the left of the view. The specified point is selected as the first point of the section plane.

   When you move the cursor toward right, the symbol of the perpendicular constraint is attached to the cursor which confirms that the line defining the section view is horizontal. This symbol also indicates that the line is normal to the extreme left vertical edge of the base view. This perpendicular constraint is applied because you snapped to the midpoint of the left vertical edge of the base view.

5. Move the cursor horizontally toward the right of the view. You will notice that a horizontal line is drawn. Move the cursor on the right of the extreme right vertical edge of the base view. Make sure that the cursor does not snap to the midpoint of the right vertical edge and the line drawn is horizontal.

6. Specify a point on the right of the right vertical edge of the base view. This point is selected as the second point of the section plane.

7. Right-click and then choose **Continue** from the Marking menu; the **Section View** dialog box is displayed and the preview of the section view attached to the cursor appears on the sheet. Also, you are prompted to specify the location of the section view. Note that hatching lines will not be displayed in the preview of the section view.

8. Specify the location of the section view below the base view, see Figure 11-4. The **Section View** dialog box is automatically closed when you specify the location of the section view.

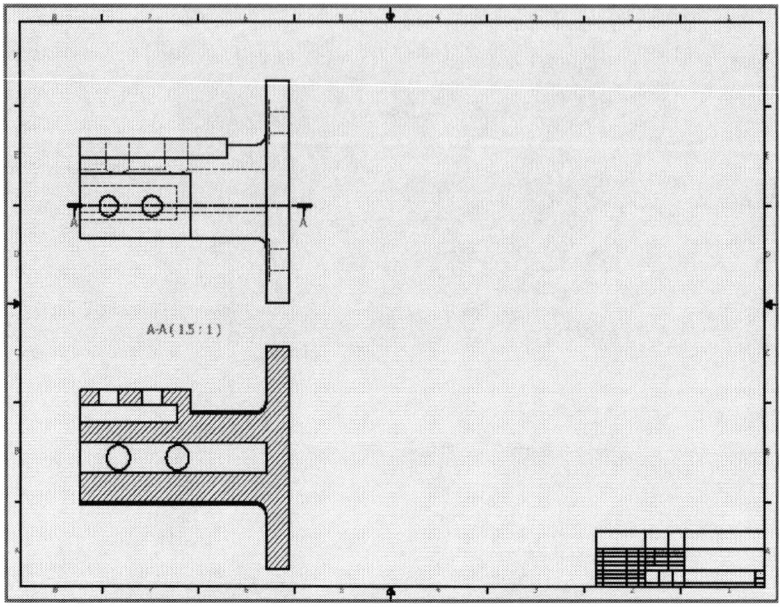

Figure 11-4 Drawing sheet with the base and section views

Generating the Isometric View of the Section View

The isometric view of the section view can be generated using the shortcut menu.

1. Move the cursor over the section view and right-click when the dotted rectangle is displayed; a shortcut menu is displayed.

2. Choose **Create View > Projected View** from the shortcut menu; you are prompted to select the view location.

3. Move the cursor toward the right of the section view and then move it upward until the preview of the isometric view appears. Now, specify the location of the view. Right-click on the view; a Marking menu is displayed. Choose **Create** from the Marking menu; the isometric view of the model is generated, as shown in Figure 11-5.

4. Save the drawing file with the name *Tutorial1.idw* at the location *C:\Inventor_2020\c11* and close it.

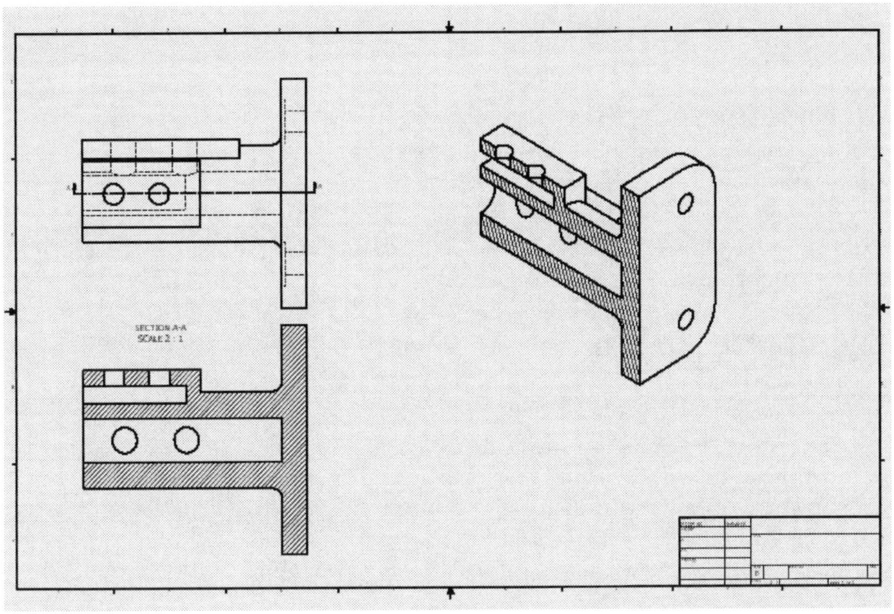

Figure 11-5 *Drawing sheet after generating the isometric view of the model*

Tutorial 2

In this tutorial, you will generate the top view, full sectioned front view, and isometric view of the section view of the Plummer Block assembly created in Tutorial 2 of the *c09* folder. The Nuts and Bolts should be excluded from the section view. Also, all sectioned components should have different hatch patterns. Use the JIS standard drawing file for generating the drawing views of the assembly. **(Expected time: 45 min)**

The following steps are required to complete this tutorial:

a. Copy the *Plummer Block* folder from the *c09* folder to the *c11* folder.
b. Generate the top view of the assembly.
c. Show the contents of the base view and then exclude the Bolts, Nuts, and Lock Nuts so that they are not sectioned.
d. Use the top view as the parent view to generate the full section front view.
e. Modify the hatch of Casting and Cap.
f. Generate the projected isometric view of the sectioned front view.

Copying the Plummer Block Folder

As mentioned earlier, you will have to copy the file that will be used to generate drawing views in the current folder. To generate the drawing views of the assembly in this tutorial, you will have to copy the folder in which the files of the assembly are stored.

1. Copy the Plummer Block folder from the location *C:\Inventor_2020\c09* to the location *C:\Inventor_2020\c11*.

Starting a New Drawing File

1. Choose the **New** tool from the **Launch** panel of the **Get Started** tab to invoke the **Create New File** dialog box.

2. In this dialog box, choose the **Metric** tab and then double-click on the **JIS.idw** option to open a JIS standard drawing file.

Generating the Top View of the Assembly

1. Choose the **Base** tool from the **Create** panel of the **Place Views** tab; the **Drawing View** dialog box is displayed.

2. In the **Component** tab of this dialog box, choose the **Open an existing file** button on the right of the **File** drop-down list; the **Open** dialog box is displayed.

3. Browse to the *Plummer Block* folder at the location *C:\Inventor_2020\c11* and then select the *Plummer Block.iam* file from it.

4. Next, choose the **Open** button to select the assembly for generating the drawing views; the preview of the view of the assembly is displayed in the graphics window.

5. Change orientation of view as top view by using ViewCube in the graphics window, as shown in Figure 11-6.

6. Modify the view scale to **1.25 : 1** in the **Scale** edit box.

   Make sure that **Hidden Line Removed** button is chosen from the **Style** area in the **Component** tab of the **Drawing View** dialog box.

7. Next, choose the **OK** button to exit the **Drawing View** dialog box; the drawing view is created.

8. Select and drag the drawing view close to the top left corner of the sheet, refer to Figure 11-6.

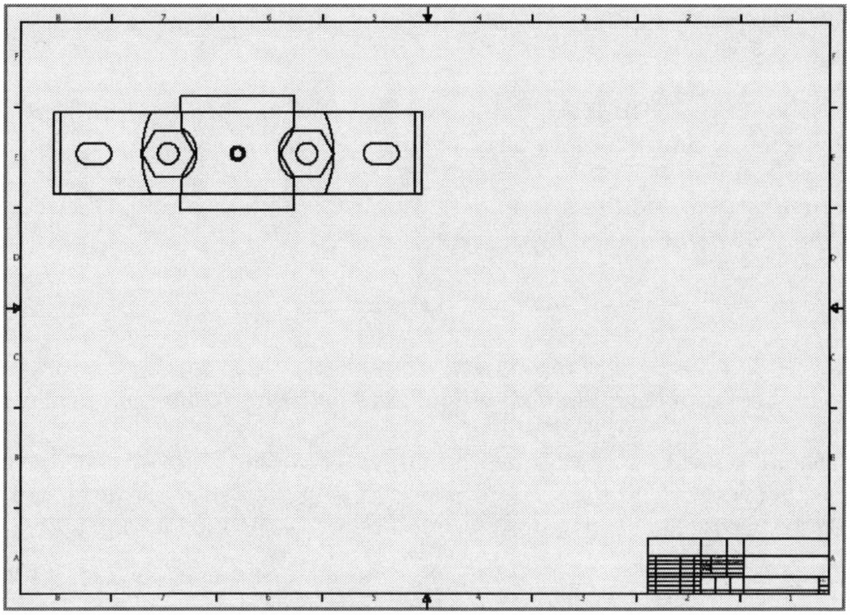

Figure 11-6 Top view of the assembly

Excluding Components from the Section Views

As mentioned in the tutorial description, the Nuts, Bolts, and Lock Nuts need to be excluded from the section view. Therefore, you need to exclude these components such that they are not sectioned in the section view.

1. Click on the + located on the left of the view node in the **Browser Bar** to display the *Plummer Block.iam* assembly. Also, a + is displayed on the left of this assembly in the **Browser Bar**.

2. Click on the + located on the left of **Plummer Block.iam** to display all components of the assembly. Press and hold the CTRL key and select both instances of Nut, Bolt, and Lock Nut using the left mouse button. The selected components are displayed in blue background in the **Browser Bar**.

3. Right-click on any selected component in the **Browser Bar**; a shortcut menu is displayed. Choose **Section Participation > None** from the shortcut menu.

4. Click anywhere on the sheet to clear the selection of components.

Generating the Section View

Since you have turned off the option for sectioning some of the components, they will not be sectioned while generating the section view.

1. Choose the **Section** tool from the **Create** panel of the **Place Views** tab; you are prompted to select the view to be sectioned.

2. Select the top view from the Graphics window; the cursor turns into a sketch cursor and you are prompted to enter the endpoints of the section line.

3. Move the cursor close to the midpoint of the extreme left vertical edge of the top view; the cursor snaps to the midpoint and turns green.

4. When the cursor snaps to the midpoint, move it horizontally toward the left and specify a point as the start point of the section plane.

5. Now, move the cursor horizontally toward the right.

6. Specify a point on the right of the extreme right vertical edge of the top view as the second point of the section plane. Note that the line should be horizontal, not inclined.

7. Right-click and then choose **Continue** from the Marking menu to display the **Section View** dialog box. The preview of the section view attached to the cursor is displayed and you are prompted to specify the location of the section view. Specify the location below the top view, as shown in Figure 11-7.

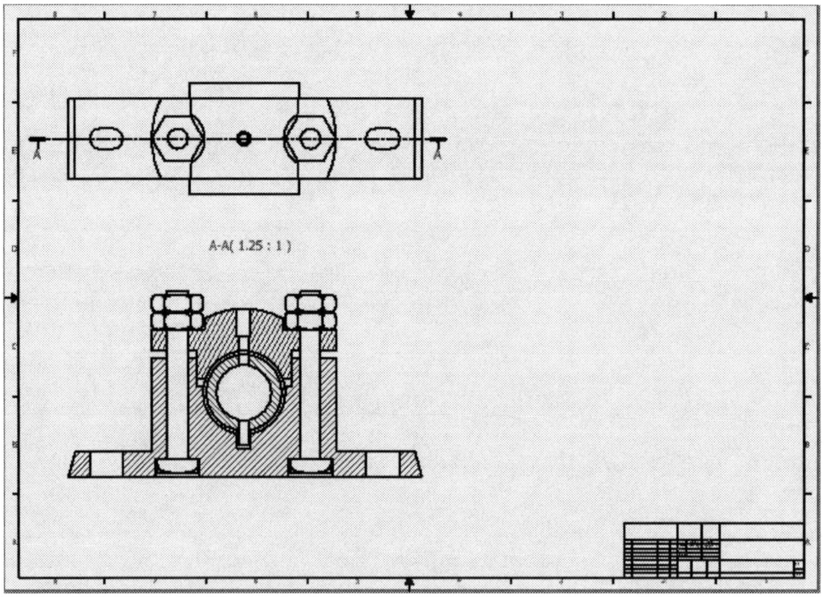

Figure 11-7 *Sheet with the top view and the sectioned front view*

Modifying Hatch Patterns

The section view displays three components in section: Casting, Cap, and Brasses. One of these components can retain the current hatching style and the hatching style in the remaining two components will be modified. In this tutorial, Brasses will retain the current style and you will modify the hatching in Casting and Cap.

1. Move the cursor over the hatching in Casting; the hatching lines turn red. Next, right-click and choose **Edit** from the shortcut menu to display the **Edit Hatch Pattern** dialog box.

2. In this dialog box, select **ISO02W100** from the **Pattern** drop-down list and then select the **Double** check box. Choose **OK** to exit this dialog box; the hatching style of the selected component is modified.

3. Move the cursor over the hatching in Cap; the hatching lines turn red. Next, right-click to display the shortcut menu. Choose **Edit** from the shortcut menu to display the **Edit Hatch Pattern** dialog box.

4. Select the **Double** check box and then choose **OK** to exit this dialog box. All three components that are sectioned have different hatch patterns now.

Generating the Isometric View of the Section View

The third view that you need to generate is the isometric view of the section view. This view is generated using the shortcut menu.

1. Move the cursor on the section view to display the bounding box. Note that the cursor should not be over any hatch pattern. When the bounding box is displayed, right-click to display the shortcut menu. Choose **Create View > Projected View** from the shortcut menu; the preview of the projected view is attached to the cursor.

2. Move the cursor toward the right of the section view in the horizontal direction and then move the cursor upward until the preview of the isometric view attached to the cursor is displayed. When the isometric view is displayed, click to specify the point to define the location of this view.

3. Right-click and then choose **Create** from the Marking menu; the isometric view of the section view is generated. The drawing sheet with all drawing views is shown in Figure 11-8.

4. Save the drawing sheet with the name *Tutorial2.idw* at the location below and then close the file.

   C:\Inventor_2020\c11\Plummer Block

   The file is saved in the Plummer Block folder because the Plummer Block assembly file that was used to generate drawing views is stored in it.

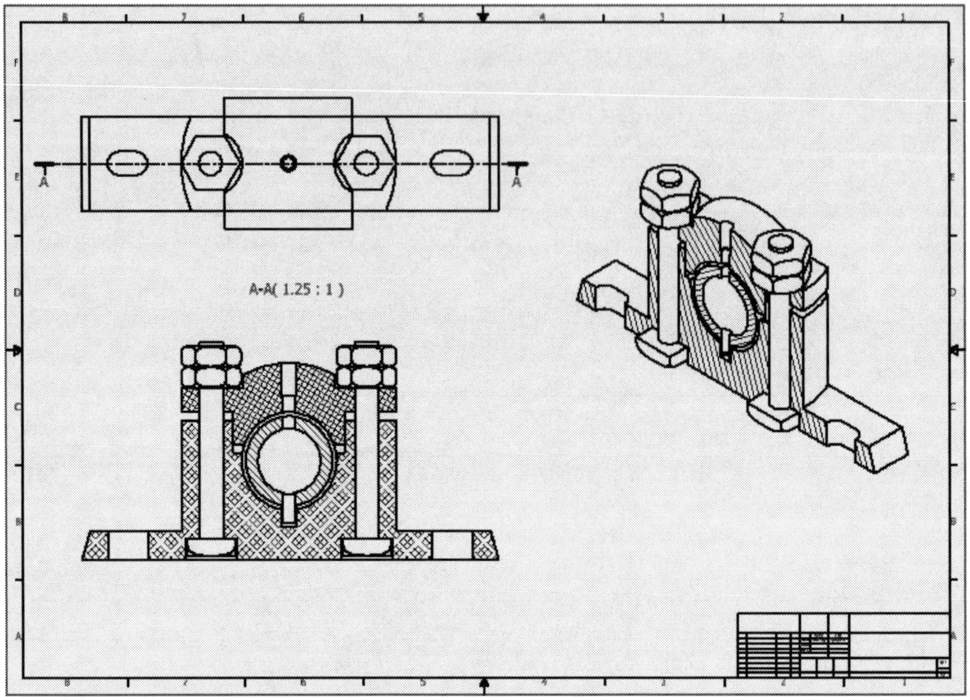

***Figure 11-8** Drawing sheet after generating all three views*

EXERCISE

Exercise 1

Generate the top view, right half sectioned front view, isometric view, and overlay view of the Double Bearing assembly with a scale of 2.5:1. Note that the overlay view should be created using positional reference. This assembly was created in Tutorial *3* of the *c10* folder. The Nut that is intersected by the cutting plane should not be sectioned and the components should have different hatch patterns, as shown in Figure 11-9. Use the JIS standards for generating the views. **(Expected time: 45 min)**

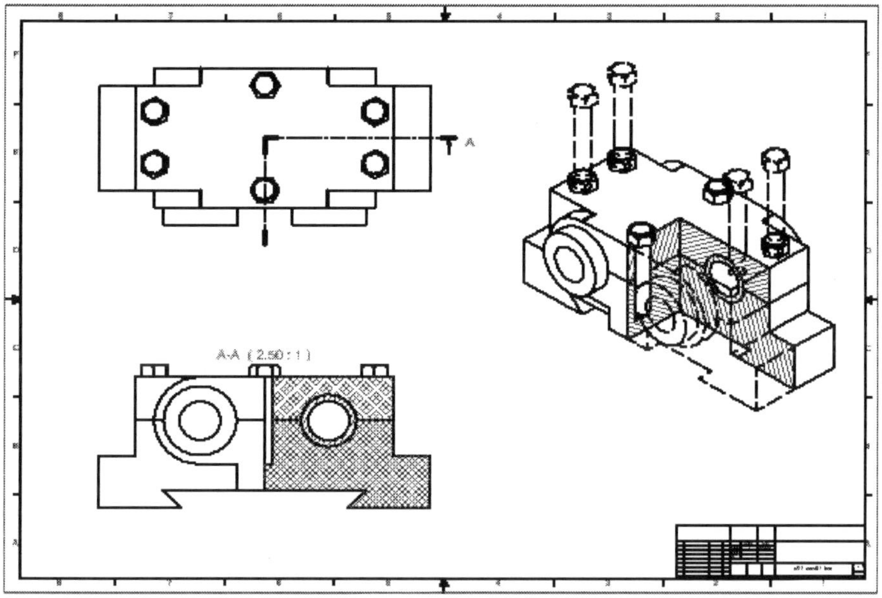

Figure 11-9 *Drawing views to be generated for Exercise 1*

This page is intentionally left blank

Chapter 12

Working with Drawing Views-II

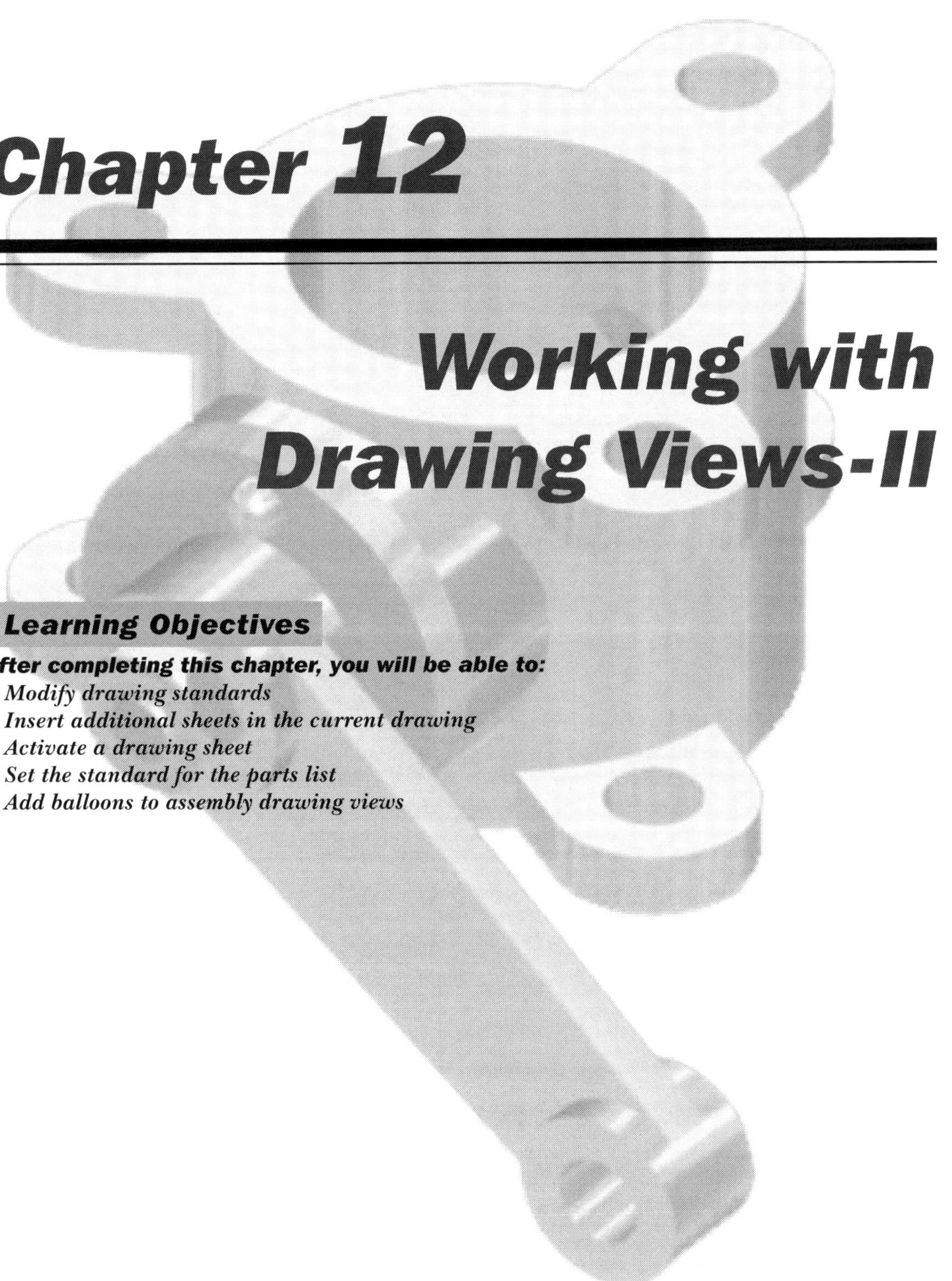

Learning Objectives

After completing this chapter, you will be able to:

- *Modify drawing standards*
- *Insert additional sheets in the current drawing*
- *Activate a drawing sheet*
- *Set the standard for the parts list*
- *Add balloons to assembly drawing views*

MODIFYING DRAWING STANDARDS

As mentioned in Chapter 11, by default, a selected sheet follows its standards in generating and dimensioning the drawing views. However, you can modify the standards of the current sheet. For example, you can open a JIS standard drawing file and assign the ANSI standards to it such that when you generate the drawing views and dimension them, the ANSI drafting standards are followed. You can modify the standards of the current sheet by choosing the **Styles Editor** tool from the **Styles and Standards** panel of the **Manage** tab.

INSERTING ADDITIONAL SHEETS INTO DRAWING

Ribbon: Place Views > Sheets > New Sheet

When you open a new drawing file, only one sheet is available. However, you can insert more drawing sheets for generating the drawing views using the **New Sheet** tool. You can also insert a new drawing sheet by using the **Browser Bar**.

ACTIVATING A DRAWING SHEET

You can activate any drawing sheet by right-clicking on it in the **Browser Bar** and then choosing **Activate** from the shortcut menu. Note that if a sheet has already been activated, this option will not be available when you right-click on a sheet in the **Browser Bar**. You can also make a sheet active by double-clicking on it in the **Browser Bar**.

DISPLAYING DIMENSIONS IN DRAWING VIEWS

As mentioned in Chapter 11, you can display the model dimensions on the drawing views while generating them. Model dimensions are also called parametric dimensions and are the dimensions that were used to create the model in the part file. These are the dimensions that were applied on the sketches or in various dialog boxes while defining features. To display model dimensions while generating a drawing view, select the **All Model Dimensions** check box in the **Recovery Options** tab of the **Drawing View** dialog box. Note that this option will not be available when you generate the drawing views of an assembly.

Retrieving Parametric Dimensions in Drawing Views

Ribbon: Annotate > Retrieve > Retrieve Model Annotations

The **Retrieve Model Annotation** tool is used to retrieve the annotation dimensions of a model after placing the drawing view.

Adding Reference Dimensions

Ribbon: Annotate > Dimension > Dimension

Autodesk Inventor allows you to add reference dimensions to a drawing view. You can do so by using the **Dimension** tool of the drafting environment. The function of this tool is similar to that of the **Dimension** tool in the **Part** module.

MODIFYING THE MODEL DIMENSIONS

Autodesk Inventor allows you to modify model dimensions displayed in a drawing view. However, as mentioned earlier, all modules of Autodesk Inventor are bidirectionally associative. This nature of Autodesk Inventor ensures that if you modify a dimension value in the **Drawing** module, the modifications will reflect on the model in the **Part** module. Therefore, you need to be very careful while modifying the model dimensions.

ADDING THE PARTS LIST

Ribbon: Annotate > Table > Parts List

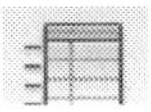

The parts list is a table, which provides information about the items, quantity, and other related description of components in an assembly. It is extremely useful for providing information related to the components of an assembly in the drawing views.

SETTING THE STANDARD FOR THE PARTS LIST

You can set the standard for the parts list using the **Style and Standard Editor [Library - Read Only]** dialog box. This dialog box is invoked when you choose the **Styles Editor** tool from the **Styles and Standards** panel of the **Manage** tab of the **Ribbon**. After invoking this dialog box, expand the **Parts List** option and then select the required parts list standard from it; the options related to the selected parts list standard will be displayed in the right pane. Using these options, you can set the parameters related to the parts list. After making the necessary modifications in the parts list standards, choose the **Save and Close** button. You will notice that the changes are reflected in the parts list on the sheet.

ADDING BALLOONS TO ASSEMBLY DRAWING VIEWS

Whenever you add the parts list to the assembly drawing views, all components in the assembly are listed in the parts list in a tabular form. You will notice that each component of the part list is assigned a unique number. As a result, if an assembly has ten components, all of them will be listed in the parts list with a different serial number assigned to them. However, in the drawing views, there is no reference about these components. Therefore, if you are not familiar with the names of the components, it is difficult to recognize them in the drawing view. To avoid this confusion, Autodesk Inventor allows you to add callouts, called balloons, to the components in the drawing view. These callouts are based on the serial number of the components in the parts list. If the serial number 1 is assigned to the component in the parts list, the callout will also show number 1. Balloons make it convenient to relate the components in the parts list to those in the drawing view. You can add balloons to the selected components manually or automatically. The methods of adding balloons are discussed next.

ADDING TEXT TO A DRAWING SHEET

Autodesk Inventor allows you to add user-defined text to a drawing sheet. Depending upon your requirement, you can add multiline text with or without a leader. The methods of adding both types of text are discussed next.

TUTORIALS

Tutorial 1

In this tutorial, you will open the drawing views of the Double Bearing assembly generated in Exercise 1 of chapter 11 and saved in *c11* folder. The drawing views were generated with positional representation. But, in this tutorial, you will create the parts list of the drawing views without positional representation. After opening the drawing views, you will add the parts list and balloons to components. Note that you will add balloons to the isometric view of the sectioned front view. The final parts list should appear as shown in Figure 12-1.

(Expected time: 45 min)

Parts List			
ITEM	QTY	NAME	DESCRIPTION
1	1	Base	Bronze
2	1	Cap	Steel
3	2	Bushing	Steel
4	6	Bolt	.50–13UNC X 4.00

Figure 12-1 *Parts list*

The following steps are required to complete this tutorial:

a. Copy the *Double Bearing* folder from the *C:\Inventor_2020\c11* folder to the current folder.
b. Open the *Exercise1.idw* file in this folder.
c. Place the default parts list by using the **Parts List** tool. Use the isometric view of the sectioned front view for placing the parts list.
d. Modify the parts list such that it appears as the one shown in Figure 12-1.
e. Add balloons to the components in the isometric view by using the **Balloon** tool.

Copying the Double Bearing Folder to the Current Folder

1. Copy the *Double Bearing* folder from *C:\Inventor_2020\c11* to the current folder.

2. Open the *Exercise1.idw* file from the location *C:\Inventor_2020\c12*.

   The drawing file is opened with the top view, sectioned front view, and isometric view generated in Exercise 1 of Chapter 11.

Note

To complete this tutorial, you must delete the positional representation of view-2 (isometric view) from the ***Browser Bar****. To do so, select the positional representation from the* ***Browser Bar*** *and right click on it. Then choose the* ***Delete*** *option from the shortcut menu displayed. Next, choose the* ***OK*** *button. You will notice that the position representation is deleted from the graphics window as well as the* ***Browser Bar.***

Placing the Parts List

As mentioned earlier, the parts list is placed using the **Parts List** tool. But, when you place the parts list, the data will be listed in it using the default parameters. For example, the fields under the **DESCRIPTION** column do not display any data. You need to modify the parts list after placing it so that it appears as the one shown in Figure 12-1.

1. Choose the **Parts List** tool from the **Table** panel of the **Annotate** tab; the **Parts List** dialog box is displayed.

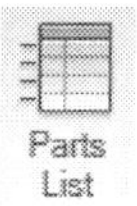

As mentioned earlier, the parts list can be placed taking the reference of a drawing view. It is recommended that the drawing view that is used as a reference for placing the parts list should have all components. On doing so, all components are listed in the parts list.

2. Select the isometric view as the reference view for placing the parts list.

3. Accept the other default options in this dialog box and choose the **OK** button; the **BOM View Disabled** dialog box is displayed. Choose the **OK** button from it; a rectangle, which is actually the parts list, gets attached to the cursor and you are prompted to specify the location of the parts list.

4. Specify the location of the parts list at the lower right corner of the sheet above the title block. The sheet with the default parts list is shown in Figure 12-2.

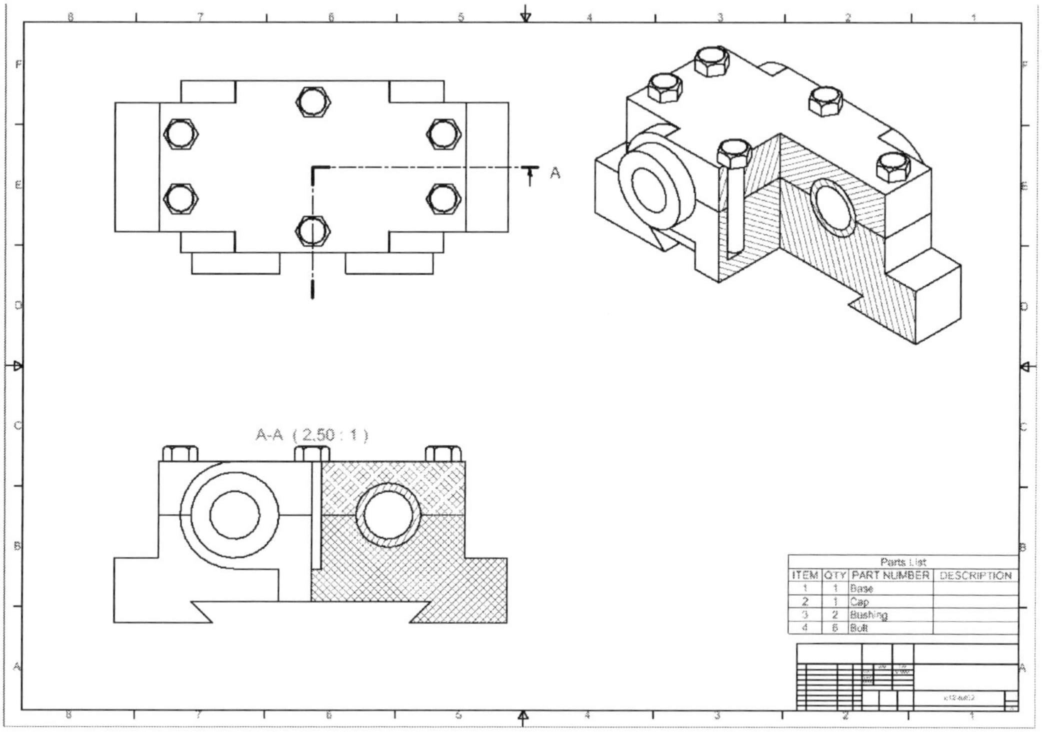

Figure 12-2 *Drawing sheet with the default parts list*

Modifying the Parts List

When you place a parts list in the drawing views, it is displayed in the drawing sheet and in the **Browser Bar**. You need to modify the parts list by changing the heading **PART NUMBER** to **NAME**. Also, you need to enter data in the fields below the **DESCRIPTION** column and center-align the data in this column.

1. Double-click on the parts list in the drawing sheet; the **Parts List** dialog box is displayed.

2. Click on the first field below the **DESCRIPTION** column and type **Bronze**.

3. Similarly, click on the remaining fields in the **DESCRIPTION** column and enter description about the remaining components. For more information about the data to be entered, refer to Figure 12-1.

   By default, the data in the **DESCRIPTION** column is left-aligned. You need to modify the alignment and make the text center-aligned.

4. Move the cursor over the heading **DESCRIPTION**. You will notice that the cursor is replaced with an arrow pointing downward.

5. Right-click and then choose **Format Column** from the shortcut menu; the **Format Column : DESCRIPTION** dialog box is displayed.

6. Choose the **Center** button on the right of **Value** in the **Justification** area to center-align the data in the fields below the **DESCRIPTION** heading in the **Part List** dialog box. Choose **OK** to exit this dialog box.

   You will notice that the data in the selected field is center-aligned.

7. Right-click on the **DESCRIPTION** heading again and then choose the **Column Width** option from the shortcut menu. Enter **60** in the **Column Width** edit box and choose **OK**. This increases the width of the fields below the **DESCRIPTION** heading.

   By default, the heading of the column that displays the name of the components is **PART NUMBER**. You need to modify this heading to **NAME**.

8. Move the cursor over the heading **PART NUMBER** and right-click when the cursor is replaced with an arrow. Next, choose **Format Column** from the shortcut menu; the **Format Column : PART NUMBER** dialog box is displayed.

9. Enter **NAME** in the **Heading** edit box and choose **OK** to exit the **Format Column : PART NUMBER** dialog box.

10. Next, choose the **OK** button to exit the **Parts List** dialog box. The sheet after editing the parts list is shown in Figure 12-3.

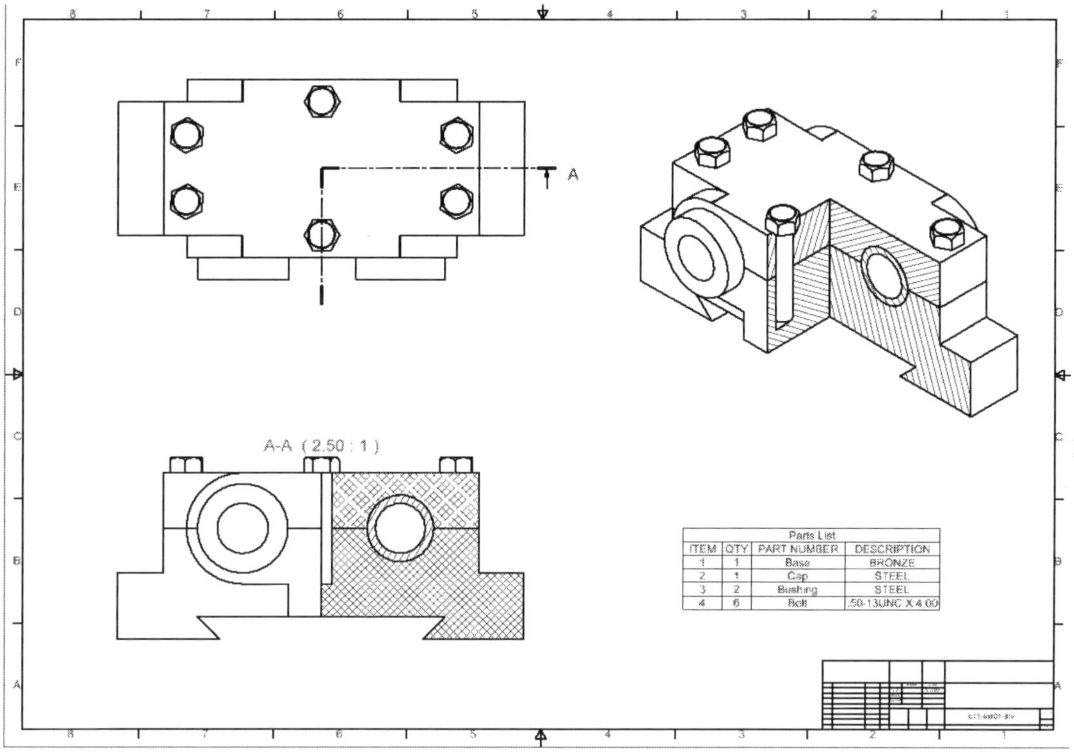

Figure 12-3 Drawing sheet after modifying the parts list

Adding Balloons to the Components

As mentioned earlier, balloons are the callouts that are attached to the components in the drawing view so that they can be referred to in the parts list. These balloons are based on the item numbers in the parts list. You can add balloons using the **Balloon** tool or the **Auto Balloon** tool. In this tutorial, balloons are added using the **Balloon** tool.

Before you add balloons, you need to set the parameters related to them.

1. Invoke the **Style and Standard Editor [Library - Read Only]** dialog box from the **Style and Standards** panel of the **Manage** tab and then expand the **Balloon** option in the left pane.
2. Select the **Balloon (JIS)** option from the left pane; the parameters related to this balloon standard are displayed in the right pane of the dialog box.
3. Choose the **Edit Leader Style** button on the right of the **Leader Style** drop-down list.
4. Select the **Filled** option from the **Arrowhead** drop-down list in the **Terminator** area.
5. Enter **6** in the **Size (X)** edit box and **2** in the **Height (Y)** edit box in the **Terminator** area. Save the changes and then exit the dialog box.

6. Choose the **Balloon** tool from **Annotate > Table > Balloon** drop-down; you are prompted to select a component. Move the cursor over one of the edges of the Bolt at the upper right corner of the assembly in the isometric view; the component is highlighted and a + sign is displayed on the left of the cursor.

7. Select the bolt; the start point of the balloon is attached to the selected edge of the bolt and the other end of the balloon is attached to the cursor.

> **Tip**
> *If you have selected a wrong component for adding the balloon, you can deselect it from the current selection set before choosing **Continue** from the shortcut menu. To do so, right-click and then choose **Back** from the shortcut menu.*

8. Specify the location of the other end of the balloon above the view, refer to Figure 12-4.

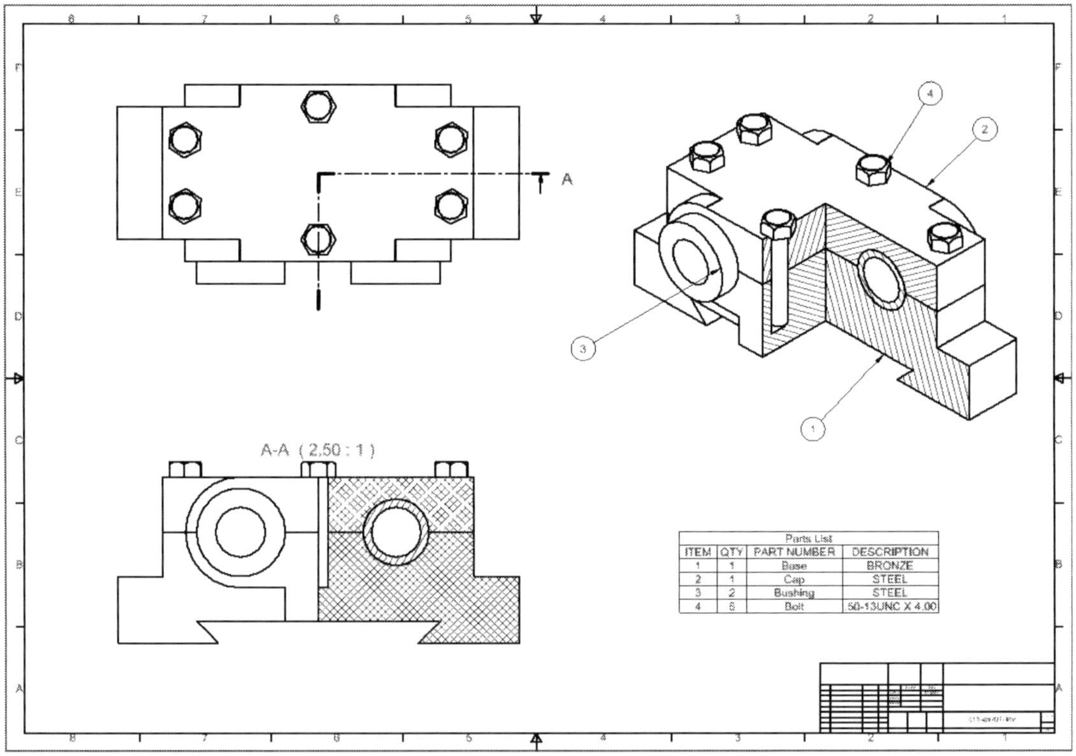

Figure 12-4 *Drawing sheet after adding balloons*

9. Now, right-click and then choose **Continue** from the shortcut menu; a balloon is created and the number 4 is displayed inside the circle. Note that number 4 corresponds to the Bolt in the parts list.

10. Next, move the cursor over the circular edge of the Bushing, which is not sectioned in the isometric view. Select it when it is highlighted; one end of the balloon is attached to the edge.

11. Specify the location of the other end of the balloon on the left of the view, refer to Figure 12-4. Right-click to display the shortcut menu and then choose **Continue** to place the balloon.

12. Similarly, add balloons to the Base and the Cap, refer to Figure 12-4.

13. After placing balloons, right-click to display the shortcut menu. Choose **Cancel [ESC]** from it to exit this tool. The drawing sheet after adding the parts list and balloons is shown in Figure 12-4.

Tip

If you double-click on the parts list after adding balloons to components, you will notice that the symbols of the balloon are displayed in front of all components in the ***Parts List*** *dialog box. These symbols suggest that the balloons corresponding to the components are added to the drawing sheet.*

To change the arrowhead of a balloon, right-click on it and then choose ***Edit Arrowhead;*** *the* ***Change Arrowhead*** *dialog box will be displayed with a drop-down list. Now, you can select the required arrowhead style from this drop-down list.*

14. Save the file with the name *Tutorial1.idw* at the location *C:\Inventor_2020\c12\Double Bearing* and then close the file.

Note

If the drawing file consists of more than one sheet, irrespective of which sheet was active while closing the file, the first sheet will be active when you open the drawing file next time.

Tutorial 2

In this tutorial, you will generate the drawing views of the Drill Press Vice assembly created in Exercise 1 of Chapter 9. The drawing views that need to be generated are shown in Figure 12-5. The parts list should appear as the one shown in Figure 12-6. You will use the ANSI mm standard sheet and the A3 size sheet for generating the drawing views. **(Expected time: 45 min)**

The following steps are required to complete this tutorial:

a. Copy the *Drill Press Vice* folder from the *c09* folder to the *c12* folder. Start a new ANSI mm standard drawing file using the **Metric** tab of the **Create New File** dialog box.
b. Modify the sheet to the A3 size sheet.
c. Modify the drafting standards and generate the required drawing views.
d. Add the parts list and modify it such that it resembles the one shown in Figure 12-6.
e. Add balloons to components.

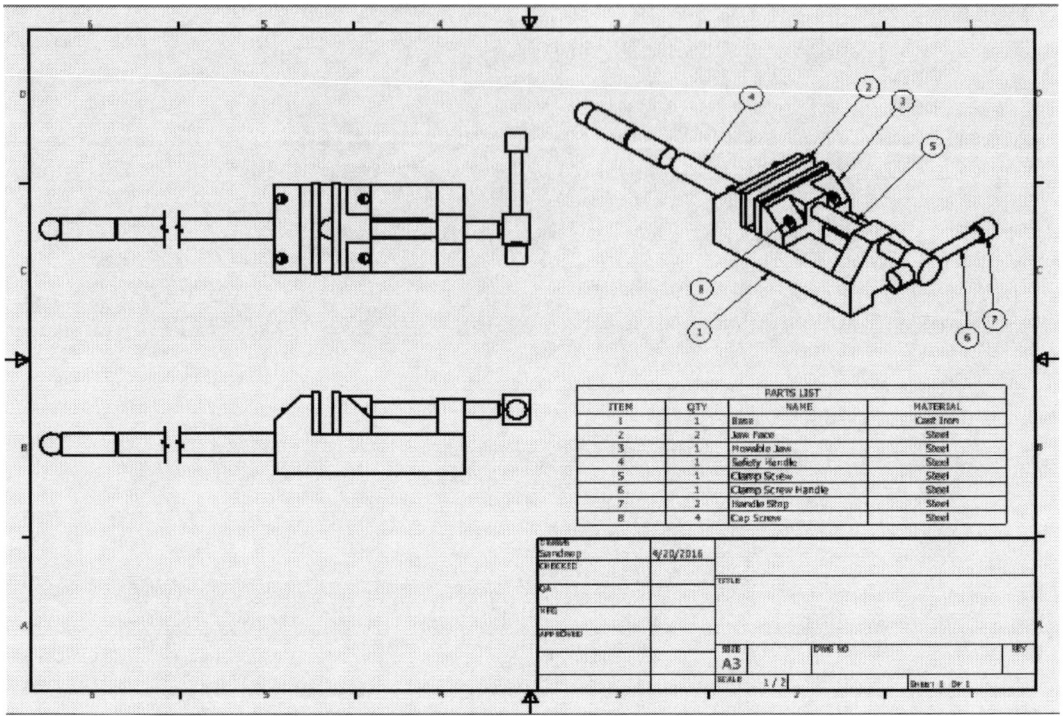

Figure 12-5 Drawing sheet for Tutorial 2

Parts List			
ITEM	QTY	NAME	MATERIAL
1	1	Base	Cast Iron
2	1	Safety Handle	Steel
3	2	Jaw Face	Steel
4	1	Movable Jaw	Steel
5	4	Cap Screw	Steel
6	1	Clamp Screw	Steel
7	1	Clamp Screw Handle	Steel
8	2	Handle Stop	Steel

Figure 12-6 Parts list to be added

Copying the Drill Press Vice Assembly

1. Copy the *Drill Press Vice* folder from the location *C:\Inventor_2020\c09* to the *c12* folder.

Starting a New ANSI mm Standard File

1. Start a new metric file with ANSI mm standards.

   The default ANSI mm standard drawing sheet is displayed. The size of the default sheet is D. You need to change this size to A3.

2. Right-click on **Sheet:1** in the **Browser Bar** and then choose **Edit Sheet** from the shortcut menu; the **Edit Sheet** dialog box is displayed.

3. Select **A3** from the **Size** drop-down list in the **Format** area. Choose **OK** to exit the dialog box; the sheet size changes to A3.

Generating the Drawing Views

1. Generate the top view of the Drill Press Vice assembly with a scale of 0.5. Break the view such that the length of the Safety Handle is reduced.

2. Generate the front and isometric views, as shown in Figure 12-7. Note that if the orientation of the views is not the same shown in Figure 12-7 then make the top face of the assembly as top in the ViewCube of the assembly environment. To do so, click on the down arrow available next to the ViewCube; a flyout is displayed. Next, choose the **Set Current View as> Top** from the flyout.

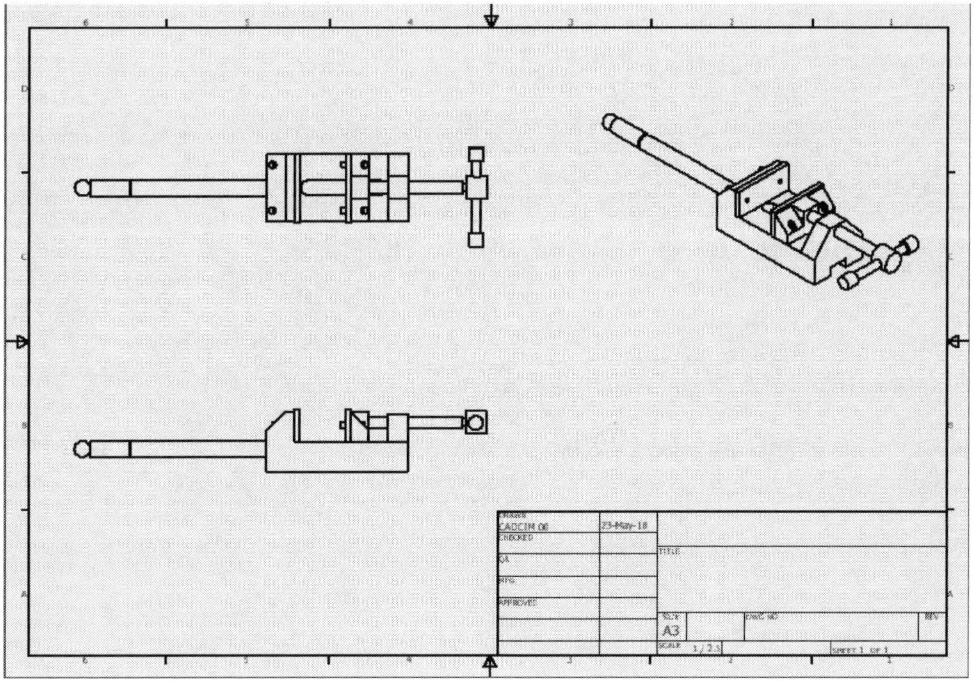

Figure 12-7 *Drawing sheet after generating the drawing views*

Placing the Parts List

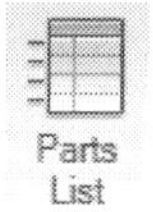

1. Choose the **Parts List** tool from the **Table** panel of the **Annotate** tab; the **Parts List** dialog box is displayed.

2. Select the isometric view as the reference view for placing the parts list.

3. Accept the default options in this dialog box and choose **OK**; the **BOM View Disabled** dialog box is displayed. Choose the **OK** button; a rectangle, which is actually the parts list, gets attached to the cursor and you are prompted to specify the location of the parts list.

4. Place the parts list above the title block.

Modifying the Parts List

1. Double-click on the parts list to display the **Parts List** dialog box.

2. Right-click on the **DESCRIPTION** heading and then choose the **Format Column** option from the shortcut menu; the **Format Column : DESCRIPTION** dialog box is displayed. Enter **MATERIAL** as the heading of this column in the **Heading** edit box.

3. Choose the **Center** button on the right of **Value** in the **Justification** area to center-align the data in the **MATERIAL** column. Next, choose **OK** to exit this dialog box.

4. Similarly, right-click on the **PART NUMBER** column and modify its heading to **NAME** in the **Heading** edit box of the **Format Column : PART NUMBER** dialog box. Next, choose **OK** to exit the dialog box.

5. Enter data in the **MATERIAL** field, based on the parts list shown in Figure 12-6. Choose **OK** to exit the dialog box.

Adding Balloons to the Components

The final step in this tutorial is to add balloons to the components in the isometric view. In this assembly, you will add balloons using the **Balloon** tool. Also, you need to drag balloons such that they are placed at proper locations in the drawing sheet. But before generating balloons, you need to modify the balloon style.

1. Invoke the **Style and Standard Editor [Library - Read Only]** dialog box and then expand the **Balloon** option in the left pane.

2. Select the **Balloon (ANSI)** option; the parameters related to the balloon standard are displayed on the right pane of the dialog box.

3. Choose the **Edit Leader Style** button on the right of the **Leader Style** drop-down list to display the leader parameters.

4. Enter **4** in the **Size (X)** edit box and **1.5** in the **Height (Y)** edit box. Choose **Save** and then **Cancel** to exit this dialog box; the size of arrowheads in the balloons is increased.

5. Choose the **Balloon** tool from **Annotate > Table > Balloon** drop-down; you are prompted to select a component.

6. Move the cursor over one of the edges of the Base in the isometric view and click to add the balloon. Next, move the cursor away from the Base and place it below the component, refer to Figure 12-8. Right-click and then choose **Continue** from the shortcut menu.

7. Similarly, add balloons to the remaining components, refer to Figure 12-8.

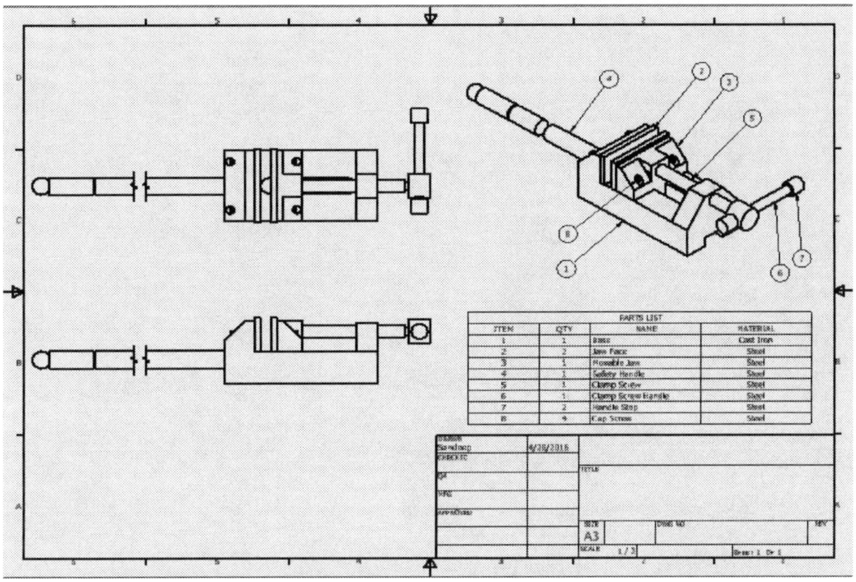

Figure 12-8 *Final drawing sheet for Tutorial 3*

8. Drag balloons to a proper location in the drawing sheet. The final drawing sheet after adding the parts list and balloons is shown in Figure 12-8.

9. Save this file with the name *Tutorial2.idw* at the location *C:\Inventor_2020\c12\Drill Press Vice* and then close the file.

EXERCISE

Exercise 1

Add the parts list and balloons to the drawing views of the Plummer Block assembly created in Tutorial 2 of Chapter 11, as shown in Figure 12-9. The parts list to be added is shown in Figure 12-10. **(Expected time: 45 min)**

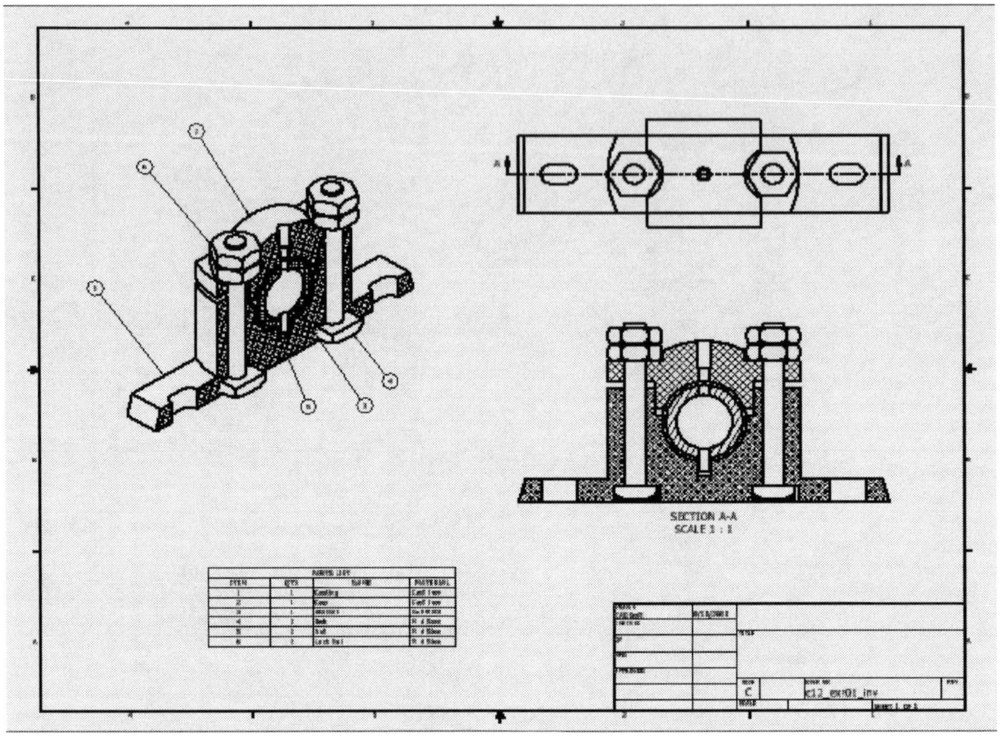

Figure 12-9 *Drawing sheet for Exercise 1*

Parts List			
ITEM	QTY	NAME	MATERIAL
1	1	Casting	Cast Iron
2	1	Brasses	Gunmetal
3	1	Cap	Mild Steel
4	2	Bolt	Mild Steel
5	2	Nut	Mild Steel
6	2	Lock Nut	Mild Steel

Figure 12-10 *Parts list for Exercise 1*

Index

L

M

N

O

P

Q

R

S

T

V

W

Z

Made in the USA
Coppell, TX
05 November 2026

11365026R00122